Research at the Intersection of the Physical and Life Sciences

Committee on Research at the Intersection of the Physical and Life Sciences

Board on Physics and Astronomy

Board on Life Sciences

Board on Chemical Sciences and Technology

Division on Engineering and Physical Sciences

Division on Earth and Life Studies

NATIONAL RESEARCH COUNCIL
OF THE NATIONAL ACADEMIES

THE NATIONAL ACADEMIES PRESS
Washington, D.C.
www.nap.edu

THE NATIONAL ACADEMIES PRESS   500 Fifth Street, N.W.   Washington, DC 20001

NOTICE: The project that is the subject of this report was approved by the Governing Board of the National Research Council, whose members are drawn from the councils of the National Academy of Sciences, the National Academy of Engineering, and the Institute of Medicine. The members of the committee responsible for the report were chosen for their special competences and with regard for appropriate balance.

This project was supported by the Department of Energy under Award No. DE-FG02-07ER46378, the National Science Foundation under Award No. CHE-0554275, the Department of Health and Human Services under Award No. N01-OD-4-2139, the Burroughs Wellcome Fund under Award No. 1007560, and the Research Corporation for Science Advancement under Award No. 7827. Any opinions, findings, conclusions, or recommendations expressed in this publication are those of the author(s) and do not necessarily reflect the views of the sponsors.

International Standard Book Number-13:   978-0-309-14751-4
International Standard Book Number-10:   0-309-14751-4
Library of Congress Control Number: 2010921434

Cover: Work at the intersection of the life sciences and the physical sciences has often been depicted in new ways of imaging or modeling biological specimens, some of which are illustrated on the cover: (1) three-dimensional distribution of membrane proteins within a cell revealed through iPALM imaging (courtesy of Harald F. Hess, Howard Hughes Medical Institute); (2) xylose isomerase crystal (courtesy of Department of Energy Office of Basic Energy Research (BER)-funded neutron Protein Crystallography Station at Los Alamos National Laboratory (LANL)); (3) simulation of confinement of DNA in viral capsid (courtesy of Molecular Dynamics and Statistical Mechanics Research Group, University of Wisconsin at Madison); (4) diffusion tension imaging of the human brain (courtesy of Thomas Schultz, University of Chicago); (5) chromosome pairs; (6) modeled structure for the enzyme D-xylose isomerase (courtesy of Department of Energy BER-funded neutron Protein Crystallography Station at LANL; (7) anglerfish ovary obtained using autofluorescence (courtesy of James E. Hayden, Wistar Institute, Philadelphia); and (8) rat cerebellum obtained using two-photon excitation fluorescence microscopy (courtesy of the National Center for Microscopy and Imaging Research at the University of California at San Diego and the National Institutes of Health.

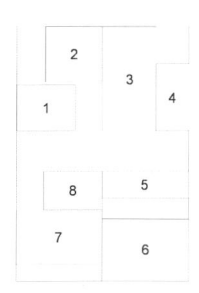

IMAGE SOURCES: (1) Harald F. Hess, Howard Hughes Medical Institute; (2) Department of Energy Office of Basic Energy Research (BER)-funded neutron Protein Crystallography Station at Los Alamos National Laboratory (LANL); (3) Molecular Dynamics and Statistical Mechanics Research Group, University of Wisconsin at Madison; (4) Thomas Schultz, University of Chicago; (6) Department of Energy BER-funded neutron Protein Crystallography Station at LANL; (7) James E. Hayden, Wistar Institute, Philadelphia; and (8) National Center for Microscopy and Imaging Research at the University of California at San Diego and the National Institutes of Health.

Copies of this report are available from: National Academies Press, 500 Fifth Street, N.W., Washington, DC 20055; (800) 624-6242 or (202) 334-3313 (in the Washington metropolitan area); Internet <http://www.nap.edu>; and the Board on Physics and Astronomy, National Research Council, 500 Fifth Street, N.W., Washington, DC 20001; Internet <http://www.national-academies.org/bpa>.

Copyright 2010 by the National Academy of Sciences. All rights reserved.

Printed in the United States of America

THE NATIONAL ACADEMIES
Advisers to the Nation on Science, Engineering, and Medicine

The **National Academy of Sciences** is a private, nonprofit, self-perpetuating society of distinguished scholars engaged in scientific and engineering research, dedicated to the furtherance of science and technology and to their use for the general welfare. Upon the authority of the charter granted to it by the Congress in 1863, the Academy has a mandate that requires it to advise the federal government on scientific and technical matters. Dr. Ralph J. Cicerone is president of the National Academy of Sciences.

The **National Academy of Engineering** was established in 1964, under the charter of the National Academy of Sciences, as a parallel organization of outstanding engineers. It is autonomous in its administration and in the selection of its members, sharing with the National Academy of Sciences the responsibility for advising the federal government. The National Academy of Engineering also sponsors engineering programs aimed at meeting national needs, encourages education and research, and recognizes the superior achievements of engineers. Dr. Charles M. Vest is president of the National Academy of Engineering.

The **Institute of Medicine** was established in 1970 by the National Academy of Sciences to secure the services of eminent members of appropriate professions in the examination of policy matters pertaining to the health of the public. The Institute acts under the responsibility given to the National Academy of Sciences by its congressional charter to be an adviser to the federal government and, upon its own initiative, to identify issues of medical care, research, and education. Dr. Harvey V. Fineberg is president of the Institute of Medicine.

The **National Research Council** was organized by the National Academy of Sciences in 1916 to associate the broad community of science and technology with the Academy's purposes of furthering knowledge and advising the federal government. Functioning in accordance with general policies determined by the Academy, the Council has become the principal operating agency of both the National Academy of Sciences and the National Academy of Engineering in providing services to the government, the public, and the scientific and engineering communities. The Council is administered jointly by both Academies and the Institute of Medicine. Dr. Ralph J. Cicerone and Dr. Charles M. Vest are chair and vice chair, respectively, of the National Research Council.

www.national-academies.org

COMMITTEE ON RESEARCH AT THE INTERSECTION OF THE PHYSICAL AND LIFE SCIENCES

ERIN K. O'SHEA, *Co-Chair,* Harvard University
PETER G. WOLYNES, *Co-Chair,* University of California at San Diego
ROBERT H. AUSTIN, Princeton University
BONNIE L. BASSLER, Princeton University
CHARLES R. CANTOR, Sequenom, Inc.
WILLIAM F. CARROLL, Occidental Chemical Corporation
THOMAS R. CECH, Howard Hughes Medical Institute
CHRISTOPHER B. FIELD, Carnegie Institution Department of Global Ecology
GRAHAM R. FLEMING, University of California at Berkeley
ROBERT J. FULL, University of California at Berkeley
SHIRLEY ANN JACKSON, Rensselaer Polytechnic Institute
LAURA L. KIESSLING, University of Wisconsin at Madison
CHARLES M. LOVETT, JR., Williams College
DIANNE NEWMAN, Massachusetts Institute of Technology
MONICA OLVERA de la CRUZ, Northwestern University
JOSÉ N. ONUCHIC, University of California at San Diego
GREGORY A. PETSKO, Brandeis University
ASTRID PRINZ, Emory University
CHARLES V. SHANK, Lawrence Berkeley National Laboratory (retired)
BORIS I. SHRAIMAN, University of California at Santa Barbara
H. EUGENE STANLEY, Boston University
GEORGE M. WHITESIDES, Harvard University

Staff

DONALD C. SHAPERO, Director, Board on Physics and Astronomy
FRANCES E. SHARPLES, Director, Board on Life Sciences
DOROTHY ZOLANDZ, Director, Board on Chemical Sciences and Technology
ADAM P. FAGEN, Senior Program Officer, Board on Life Sciences
JAMES C. LANCASTER, Program Officer, Board on Physics and Astronomy
KATHRYN J. HUGHES, Program Officer, Board on Chemical Sciences and Technology
NATALIA MELCER, Program Officer, Board on Physics and Astronomy
LAVITA COATES-FOGLE, Senior Program Assistant

SOLID STATE SCIENCES COMMITTEE

BARBARA JONES, *Chair,* IBM Almaden Research Center
MONICA OLVERA de la CRUZ, *Vice-Chair,* Northwestern University
DANIEL AROVAS, University of California at San Diego
COLLIN L. BROHOLM, The Johns Hopkins University
PAUL CHAIKIN, New York University
GEORGE CRABTREE, Argonne National Laboratory
ANDREA J. LIU, University of Pennsylvania
JOSEPH ORENSTEIN, University of California at Berkeley
ARTHUR P. RAMIREZ, LGS, a subsidiary of Alcatel-Lucent, and Columbia University
RICHARD A. REGISTER, Princeton University
MARK STILES, National Institute of Standards and Technology
DALE J. VAN HARLINGEN, University of Illinois at Urbana-Champaign
FRED WUDL, University of California at Santa Barbara

Staff

DONALD C. SHAPERO, Director, Board on Physics and Astronomy
MICHAEL H. MOLONEY, Associate Director
JAMES C. LANCASTER, Program Officer
LAVITA COATES-FOGLE, Senior Program Assistant
BETH DOLAN, Financial Associate

BOARD ON PHYSICS AND ASTRONOMY

MARC A. KASTNER, *Chair,* Massachusetts Institute of Technology
ADAM S. BURROWS, *Vice-Chair,* University of Arizona
JOANNA AIZENBERG, Harvard University
JAMES E. BRAU, University of Oregon
PHILIP H. BUCKSBAUM, Stanford University
PATRICK L. COLESTOCK, Los Alamos National Laboratory
RONALD C. DAVIDSON, Princeton University
ANDREA M. GHEZ, University of California at Los Angeles
PETER F. GREEN, University of Michigan
LAURA H. GREENE, University of Illinois at Urbana-Champaign
MARTHA P. HAYNES, Cornell University
JOSEPH HEZIR, EOP Group, Inc.
MARK KETCHEN, IBM Thomas J. Watson Research Center
ALLAN H. MacDONALD, University of Texas at Austin
PIERRE MEYSTRE, University of Arizona
HOMER A. NEAL, University of Michigan
JOSE N. ONUCHIC, University of California at San Diego
LISA J. RANDALL, Harvard University
CHARLES V. SHANK, Lawrence Berkeley National Laboratory (retired)
MICHAEL S. TURNER, University of Chicago
MICHAEL C.F. WIESCHER, University of Notre Dame

Staff

DONALD C. SHAPERO, Director
MICHAEL H. MOLONEY, Associate Director
ROBERT L. RIEMER, Senior Program Officer
JAMES C. LANCASTER, Program Officer
DAVID B. LANG, Program Officer
CARYN J. KNUTSEN, Research Associate
BETH DOLAN, Financial Associate

BOARD ON LIFE SCIENCES

KEITH YAMAMOTO, *Chair,* University of California at San Francisco
ANN M. ARVIN, Stanford University
BONNIE L. BASSLER, Princeton University
VICKI L. CHANDLER, Gordon and Betty Moore Foundation
SEAN EDDY, HHMI Janelia Farm Research Campus
MARK D. FITZSIMMONS, John D. and Catherine T. MacArthur Foundation
DAVID R. FRANZ, Midwest Research Institute
LOUIS J. GROSS, University of Tennessee at Knoxville
JO HANDELSMAN, University of Wisconsin at Madison
CATO T. LAURENCIN, University of Connecticut
JONATHAN D. MORENO, University of Pennsylvania
ROBERT M. NEREM, Georgia Institute of Technology
CAMILLE PARMESAN, University of Texas at Austin
MURIEL E. POSTON, Skidmore College
ALISON G. POWER, Cornell University
BRUCE W. STILLMAN, Cold Spring Harbor Laboratory
CYNTHIA WOLBERGER, Johns Hopkins University School of Medicine
MARY WOOLLEY, Research! America

Staff

FRANCES E. SHARPLES, Director
JO L. HUSBANDS, Scholar/Senior Project Director
ADAM P. FAGEN, Senior Program Officer
ANN H. REID, Senior Program Officer
MARILEE K. SHELTON-DAVENPORT, Senior Program Officer
INDIA HOOK-BARNARD, Program Officer
ANNA FARRAR, Financial Associate
CARL-GUSTAV ANDERSON, Senior Program Assistant
AMANDA P. CLINE, Senior Program Assistant
AMANDA MAZZAWI, Program Assistant

BOARD ON CHEMICAL SCIENCES AND TECHNOLOGY

F. FLEMING CRIM, *Co-Chair*, University of Wisconsin at Madison
GARY S. CALABRESE, *Co-Chair*, Corning, Inc.
BENJAMIN ANDERSON, Lilly Research Laboratories
PABLO G. DEBENEDETTI, Princeton University
RYAN R. DIRKX, Arkema, Inc.
MARY GALVIN-DONOGHUE, Air Products and Chemicals Materials
PAULA T. HAMMOND, Massachusetts Institute of Technology
CAROL J. HENRY, Advisor and Consultant
RIGOBERTO HERNANDEZ, Georgia Institute of Technology
CHARLES E. KOLB, Aerodyne Research, Inc.
MARTHA A. KREBS, California Energy Commission
CHARLES T. KRESGE, Dow Chemical Company
SCOTT J. MILLER, Yale University
DONALD PROSNITZ, RAND Corporation
MARK A. RATNER, Northwestern University
ERIK J. SORENSEN, Princeton University
WILLIAM C. TROGLER, University of California at San Diego
THOMAS H. UPTON, ExxonMobil

Staff

DOROTHY ZOLANDZ, Director
ANDREW CROWTHER, Postdoctoral Research Associate
KATHRYN J. HUGHES, Program Officer
TINA MASCIANGIOLI, Senior Program Officer
ERICKA McGOWAN, Associate Program Officer
JESSICA L. PULLEN, Administrative Assistant
SHEENA SIDDIQUI, Research Assistant
LYNELLE VIDALE, Project Assistant

Acknowledgment of Reviewers

This report has been reviewed in draft form by individuals chosen for their diverse perspectives and technical expertise, in accordance with procedures approved by the National Research Council's Report Review Committee. The purpose of this independent review is to provide candid and critical comments that will assist the institution in making its published report as sound as possible and to ensure that the report meets institutional standards for objectivity, evidence, and responsiveness to the study charge. The review comments and draft manuscript remain confidential to protect the integrity of the deliberative process. We wish to thank the following individuals for their review of this report:

Marlene Belfort, New York State Department of Health
Robert Dimeo, National Institute of Standards and Technology
James Heath, California Institute of Technology
Jennifer Lippincott-Schwartz, National Institutes of Health
Andrea Liu, University of Pennsylvania
Peter Moore, Yale University
Aravi Samuel, Harvard University
Philip Sharp, Massachusetts Institute of Technology
Erik Sorensen, Princeton University

Although the reviewers listed above have provided many constructive comments and suggestions, they were not asked to endorse the conclusions or recommendations, nor did they see the final draft of the report before its release. The review of this report was overseen by W. Carl Lineberger, University of Colorado at Boulder. Appointed by the National Research Council, he was responsible for making certain that an independent examination of this report was carried out in accordance with institutional procedures and that all review comments were carefully considered. Responsibility for the final content of this report rests entirely with the authoring committee and the institution.

Contents

SUMMARY 1

1 INTRODUCTION 9

2 GRAND CHALLENGES 12
   Grand Challenge 1. Synthesizing Lifelike Systems, 12
   Grand Challenge 2. Understanding the Brain, 14
   Grand Challenge 3. Predicting Individual Organisms' Characteristics
      from Their DNA Sequence, 15
   Grand Challenge 4. Interactions of the Earth, Its Climate, and the
      Biosphere, 16
   Grand Challenge 5. Understanding Biological Diversity, 18

3 SOCIETAL CHALLENGES 20
   Identifying and Combating Biological Threats, 21
      Early Detection and Intervention, 21
      Prediction of Susceptibility to Disease and Its Prevention, 22
   Climate and Its Interface with Biology, 23
      Complex Feedback Loops in Climate Science, 23
      Implications of Renewable Energy, 23
   Medicine, 24
      Imaging, 24
      Treatment and Devices, 25

Agriculture as a Resource for Food and Energy, 25
    Building Better Plants and Getting More Out of Them, 26
    Hydrogenases and Synthetic Photosynthesis, 27
    Beyond Combustion, 27
Materials Science, 28
Opportunities, 28
References, 29

4   COMMON THEMES AT THE INTERSECTION OF BIOLOGICAL AND PHYSICAL SCIENCES   30
    Interaction and Information: From Molecules to Organisms and Beyond, 30
    Dynamics, Multistability, and Stochasticity, 36
    Self-organization and Self-assembly, 42
    Conclusion, 48
    References, 49

5   ENABLING TECHNOLOGIES AND TOOLS FOR RESEARCH   51
    Introduction, 51
    Physical Basis of Molecular Recognition, 52
    Structures and Dynamics Within Cells, 53
        Cellular Environment, 53
        Interactions Within Cells, 55
        Examining Structures Within Cells, 56
    Theory and Simulations, 60
    Collective Dynamics, 64
    Complex Community Signals and Shared Resources at Large Length Scales, 65
    References, 68

6   ENABLING RESEARCH AT THE INTERSECTION: PROMOTING TRAINING, SUPPORT, AND COMMUNICATION ACROSS DISCIPLINES   69
    Connections Between Disciplines, 70
        Culture of Separation Between the Life and Physical Sciences, 70
        Culture and Organization of Academia, 72
    Organization of Support for Research, 73
    Supporting Transformative Research, 78
    Educating Scientists at the Intersection of the Physical and Life Sciences, 82

Enabling Interdisciplinary Research Starting at the Undergraduate Level, 82
Integrating Life and Physical Sciences for Graduate Students and Postdoctoral Researchers, 85
References, 89

APPENDIXES

A	Statement of Task	93
B	Meeting Agendas	94
C	Biographies of Committee Members	99

Summary

"... the action most worth watching is not at the center of things but where edges meet ... shorelines, weather fronts, international borders ..."

Anne Fadiman, *The Spirit Catches You and You Fall Down*

Almost since their inception, the natural sciences, those fields that use the scientific method to study nature, have been divided into two branches: the biological sciences and the physical sciences. In part, this division can be viewed as a convenient social contrivance. However, over time it has also served more functional purposes. Physical scientists, when seeking the fundamental laws, have found it necessary to focus on the simplest of systems—elementary particles, atoms, and molecules—items clearly not alive. It also has been convenient for biological scientists to investigate the immense diversity of living things and their elaborate inner workings without simultaneously accepting the burden of trying to follow these complexities down to their atomic level roots.

Today, while it still is convenient to classify most research in the natural sciences as either biological or physical, more and more scientists are quite deliberately and consciously addressing problems lying at the intersection of these traditional areas. This report focuses on their efforts. As directed by the charges in the statement of task (see Appendix A), the goals of the committee in preparing this report are several fold. The first goal is to provide a conceptual framework for assessing work in this area—that is, a sense of coherence for those not engaged in this research about the big objectives of the field and why it is worthy of attention from fellow scientists and programmatic focus by funding agencies. The second goal is to assess current work using that framework and to point out some of the more promising opportunities for future efforts, such as research that could significantly benefit society. The third and final goal of the report is to set out strategies for realizing those benefits—ways to enable and enhance collaboration so that the United States can take full advantage of the opportunities at this intersection.

CONCEPTUAL FRAMEWORK FOR ASSESSING THIS INTERSECTION

Any attempt to provide an all-inclusive framework for this work will inevitably leave out research that belongs within it. With that caveat, a good way to think of research at this intersection is that it turns ways of looking at things—both figuratively and literally—from their original purpose and uses them to tackle new problems, often in ways far removed from when they were first conceived.

Most—but not all—of the new problems being addressed at this intersection are biological ones, largely because of the incredible richness of this field. The realm of biology is immense, involving complicated structures as small as molecules and as large as the biosphere and timescales that range from submicroseconds to eons. Answers to these problems seek not only to describe how the individual structures, in their immense complexity and diversity, work but also how they interplay. A very rich source of potential questions indeed.

The ways of looking often come from the physical sciences. Those ways might be conceptual—approaches for looking at and solving problems—or analytical—methods for extracting understanding from data—or technical—tools for collecting information needed to address the problem at hand. But it is this intermingling of problems from one arena and ways of looking at them from another arena that makes this intersectional area between the biological and physical sciences so rich and offers many of the opportunities that reside there. The committee expects that ideas will emerge from such studies that will go well beyond the intersection and transform both the biological and physical sciences.

CURRENT WORK AT THE INTERSECTION

What, then, are some of the areas being explored at this intersection? Interestingly, many share common conceptual themes, several of which are discussed in this report. Interactions appear in both branches, albeit with much different content and contexts. Describing how individual particles interact—what forces and energy exchanges cause crystalline materials to form, and matter in all phases to display characteristic behavior and to undergo phase changes—are mainstays of the world of physics. However, these ways of thinking about and discussing how inanimate objects interact have been found useful to scientists attempting to answer questions about the interplay of biological matter at many different levels.

Another area finding fertile ground and producing fruitful cross-research opportunities centers on the dynamics of systems. Equilibrium, multistability, and stochastic behavior—concepts familiar to physicists and chemists—are now being used to tackle issues involved in living systems such as adaptation, feedback, and emergent behavior. Ideas of pattern formation that are at the heart of condensed matter physics now help us to understand biological self-assembly and the development of biological systems.

This report also discusses how some of the mysteries of the biological world have been unraveled using tools and techniques developed in the physical sciences. These tools include not only imaging devices, both photon- and matter-based, but also computational models and algorithms. While many of them are used interchangeably by the two fields, others must be modified. However, to reach the heart of biological systems, even more sophisticated investigatory technologies and tools will be needed, many of which have not even been imagined much less developed.

In preparing this report, the committee was mindful of the vastness of the number of topics that arguably comes within the ambit of this report's subject matter. Work taking place at the intersections of engineering and the life sciences and of materials development and the life sciences covers but two of such topics. Both are fascinating examples of where the meshing of different cultures and sets of ideas can produce much fruitful discussion and advancement.[1] However, the statement of task for this committee focuses on research, limiting this report to more basic activities than those typically involved in engineering and materials development. Further, the committee acknowledges that the research that is the subject matter of this report both arises from and depends upon the rich, ongoing efforts taking place within the core disciplines of the physical and life sciences. Such intersectional research serves to supplement rather than to supplant the scientific advances being made in the more traditional fields.

PROMISING OPPORTUNITIES FOR FUTURE EFFORTS

Some of the most fundamental challenges in this area and near-term prospects for successfully meeting them are discussed in the form of five Grand Challenges:

---

[1] Some have been covered in other NRC reports. See, for example, *Inspired by Biology: From Molecules to Materials to Machines*, Washington, D.C.: The National Academies Press (2008) and *A New Biology for the 21st Century*, Washington, D.C.: The National Academies Press (2009). Others might be the focus of future reports.

- *Grand Challenge 1.* Natural substances display remarkable architecture, demonstrating the immense breadth of what can be achieved in developing structures and systems. Can the skills and knowledge-sets of biological and physical scientists be combined to provide greater insight into identifying those structures, capabilities, and processes that form the basis for living systems, and then to use that insight to construct systems with some of the characteristics of life that are capable, for example, of synthesizing materials or carrying out functions as yet unseen in natural biology?
- *Grand Challenge 2.* The human brain may be nature's most complex system. Can we understand how it works and build on that understanding to predict brain function? Addressing this challenge will require drawing on the resources of the physical sciences, both existing and to be developed, from imaging techniques to modeling capabilities.
- *Grand Challenge 3.* Genes and the environment interact to produce living organisms. Can we deepen our understanding of those interactions to begin to comprehend how organisms change over time—how they age and heal, for example—and from that understanding realize the promise of personalized medicine and access to better health care?
- *Grand Challenge 4.* Earth interacts with its climate and the biosphere through strikingly different yet intertwined mechanisms that operate over vast ranges of time and space. Can life and physical scientists develop an effective approach for understanding how these mechanisms interplay and use that understanding to develop strategies that will preserve this heritage?
- *Grand Challenge 5.* Living systems display remarkable diversity, serving to protect communities from harm. This diversity is declining, however, as the result of human activities, yet efforts to understand its role in the health of a species or an ecosystem have only recently been undertaken. Can knowledge gained at the intersection of the life and physical sciences teach us how to prosper while sustaining the diversity that allows life to flourish?

Further research at this intersection not only will advance our understanding of the fundamental questions of science, but will also significantly impact public health, technology, and stewardship of the environment for the benefit of society. In the world of technology, our economy clearly is based on materials, but no synthetic material in use today is as complex as a dead piece of wood, let alone a living organism. Prospects for a material world as adaptive and robust as living things have been the stuff of science fiction for decades. To achieve these dreams requires a greater understanding of the organizing principles of life. For public health, the complexity of molecular recognition and the emergent regulation of physiology must be better understood if drug design is to progress from the art it is to engineering science. Without understanding the diversity encompassed in human biology, personalized medicine will remain more a hope than a reality.

We often think of environmental challenges as being biological ("Save the whales!") or physical ("Limit greenhouse gases!") but, again, this distinction between the disciplines is a distortion. The constant interplay between the biological sciences and the physical sciences is profound when Earth is viewed as an entire system.

STRATEGIES FOR ENABLING AND ENHANCING WORK AT THE INTERSECTION

Throughout the report, the committee recommends ways to accelerate progress in this field. Some of these recommendations are implicit. By describing the vast array of outstanding questions at this intersection, it hopes to intrigue some of its fellow scientists to venture into this area and perhaps find interesting questions to address. Of the report's explicit recommendations, several are directed to those administering the faculties and resources of our great research institutions. In both the academic and business world, the cultures of the biological and physical sciences have evolved separately. Indeed both of these broad areas maintain numerous subcultures within themselves. It might seem that the daily life of a physician and that of a professor of theoretical physics would have nothing in common, but they must appreciate each other's insights if the scientific challenges at this interface are to be met and the societal benefits realized. Just as important, educational institutions need to develop multidisciplinary research and education opportunities that transcend the traditional departmental structure. They need to establish curricula and training opportunities to prepare the next generation of scientists to grapple with the questions posed at this intersection.

The committee recognizes that the needed changes will not come about just because of this report. Federal and private funding agencies will have to establish policies and programs that provide the appropriate incentives. Professional societies will need to break down disciplinary barriers and promote scholarships at the intersection of disciplines. And academic leaders will need to take what steps they can to facilitate the changes. This report describes the many profound and societally important scientific issues yet to be explored in this area. In it, the committee hopes to make the case that inertia and the understandable resistance to change must be overcome and that necessary structural changes in academic departments and curricula need to be undertaken.

RECOMMENDATIONS

- Universities should establish science curriculum committees that include both life scientists and physical scientists to coordinate curricula between science departments and to plan introductory courses that prepare both those who would major in the life sciences and those who would enter the physical sciences.
- Professional scientific societies should partner with peer societies across the life and physical sciences to organize workshops and provide resources that will facilitate multidisciplinary education for undergraduates.
- Federal and private funding agencies should offer seed grants to academic institutions to develop new introductory courses that incorporate both the physical and life sciences and to professional societies for organizing workshops and developing resources for multidisciplinary education. They should also support research to identify best practices in such education.
- Federal and private funding agencies should offer expanded training grants that explicitly include graduate students and postdoctoral researchers from fields across the life and physical sciences and that require the involvement of academic departments from both the physical and life sciences. Funding agencies should also offer administrative supplements to existing research grants that would enable a principal investigator in the life sciences to support a postdoctoral researcher with a background in the physical sciences, or vice versa.

The report also makes recommendations to help provide better support and guidance for research in these areas. The committee hopes that this report makes the case that much of the best science at this intersection has difficulty finding a financial home and resources. All too often, the most interesting questions that researchers seek to address here do not readily fall within the purview of a particular agency program or review structure. Accordingly, the committee calls for changes in funding mechanisms that will produce effective collaboration and cooperation among federal agencies that support research in the physical and life sciences.[2] Established investigators trained in one discipline should have the opportunity to receive training in another, so they may apply their experience to multidisciplinary problems. Mechanisms need to be put in place to support investigator-initiated multidisciplinary research where review of the proposal assesses the candidate's previous work rather than just the research being proposed.

---

[2] See also the discussion in the NRC report *A New Biology for the 21st Century*, Washington, D.C.: The National Academies Press (2009).

RECOMMENDATION
The Office of Science and Technology Policy (OSTP) and the Office of Management and Budget (OMB) should develop mechanisms to ensure effective collaboration and cooperation among federal agencies that support research at the nexus of the physical and life sciences. In particular, OSTP and OMB should work with federal science agencies to establish standing mechanisms that facilitate the funding of interagency programs and coordinate the application and review procedures for such joint programs. Moreover, the National Science and Technology Council should establish a standing interagency working group on multidisciplinary research within its Committee on Science, with focus on the intersection of the physical and life sciences.

RECOMMENDATION
Federal and private funding agencies should offer opportunities for both early-career and established investigators trained in one discipline to receive training in another and apply their experience and training to interdisciplinary problems. In particular, postdoctoral career awards should be established that facilitate the transition of a candidate prepared in a physical science field to apply that training to important questions in the life sciences and vice versa. Funding agencies should also provide expanded support for experienced investigators to receive training in a new field, perhaps in the form of sabbatical fellowships.

RECOMMENDATION
Federal and private funding agencies should enhance the ability of more than one researcher to serve as principal investigator (PI) on research projects. Each PI should receive full credit for participation on the grant, with the lead PI serving as the administrative contact.

RECOMMENDATION
Federal and private finding agencies should devote a portion of their resources to support potentially transformative research, including opportunities at the intersection of the physical and life sciences. These sponsors should have peer review procedures that incorporate the viewpoints of scientists from a variety of disciplines. Moreover, they should continually assess the effectiveness of these grant programs and the review procedures to ensure that they are meeting the desired aims.

RECOMMENDATION
Federal and private funding agencies should expand support for interdisciplinary and multidisciplinary research and education centers. In particular, extramural funding should be provided to establish and maintain center infrastructure and research expenses. Initial (e.g., 5-year) salary support for investigators performing research that spans disciplines should also be included, with continuing salary support for faculty associated with the center provided by the host institution(s) or department(s). To support these centers, universities will need to implement multidepartment hiring practices and tenure policies that support faculty working collaboratively within and across multiple disciplines, establish shared resources, and provide incentives for departments to promote multidepartmental research and cross-disciplinary teaching opportunities.

Many of these recommendations are not new, but instead resemble those rendered by previous committees about the need to break away from "stove piping"—narrowly focused and isolated funding programs—and to implement new ways for evaluating funding opportunities and prioritizing funding for the most promising research. These resemblances should be seen as a renewed acknowledgement by this committee that such changes remain important and continue to be necessary to take full advantage of the research opportunities at this interface.

As noted by President Obama in his remarks to the National Academy of Sciences in April 2009, change and convergence are key to fully meeting the challenges and opportunities at this intersection:

> In biomedicine . . . we can harness the historic convergence between life sciences and physical sciences that's underway today [by] undertaking public projects—in the spirit of the Human Genome Project—to create data and capabilities that fuel discoveries in tens of thousands of laboratories and identifying and overcoming scientific and bureaucratic barriers to rapidly translate scientific breakthroughs into diagnostics and therapeutics that serve patients.[3]

New cultures must be forged and scientists must grow as comfortable in them as they are in their existing subcultures. There must be funding for work in those new cultures that extends beyond existing-culture "stove pipes." Most important, they must prepare the rising generation to mine new-culture opportunities without losing touch with scientists in the traditional disciplines or the principles of such disciplines. The future will be driven by progress at this intersection.

---

[3] Remarks by the President at the National Academy of Sciences Annual Meeting on April 27, 2009; available at http://www.whitehouse.gov/the_press_office/Remarks-by-the-President-at-the-National-Academy-of-Sciences-Annual-Meeting/. Last accessed September 3, 2009.

1

Introduction

Humans compartmentalize. We group ideas and concepts together, perhaps to help us better comprehend the world, or perhaps simply for comfort. The world of science is no exception. We tend to separate the social sciences from the natural sciences, chemistry from physics, and biology from psychology. Important for this report, the life sciences are rarely included in the same academic department as the physical sciences,[1] and it is not uncommon for these fields to be taught in separate colleges. And even within the broad categories of life sciences or physical sciences, microbiologists are likely to be in separate departments from ecologists, while physicists may rarely interact with their chemist colleagues.

To be sure, there are different approaches and methods of analysis in the different disciplines. For example, physicists are accustomed to seeking those few grand, foundational laws that describe all physical behavior and then using those laws to divulge the inner workings of the world. Constrained by these laws, and the limits of mathematical and computational capabilities, they are able to describe with rigor only systems that are, for the most part, very simple (elementary particles, atoms and molecules), very ordered (crystalline material in its many forms), or very disordered (those systems containing incomprehensibly large numbers of particles whose general characteristics and phase transformations are described using thermodynamics). As physical scientists, chemists are equally constrained

---

[1] Throughout this report, the terms "physical sciences" and "life sciences" are meant to include a variety of disciplines. For example, physical sciences includes physics, chemistry, mathematics, and related fields.

by a small number of fundamental laws. They seek to understand how small bits of matter—atoms and molecules—are internally composed, and how they absorb and transmit energy and react with each other.

In contrast, the life sciences seek to understand a natural world that is multifaceted—perhaps even messy—and almost never in the steady state. Whereas fundamental laws drive the physical sciences, diversity and complexity are the key characteristics of the life sciences. This latter world begins at the smallest scales, with the biomolecules of which all living matter is made, and then extends to cells, tissues, and organs, to complete organisms, and then to their interactions with each other and with their environments, first on a local level and then globally. Biologists have traditionally pursued their studies without feeling the need to trace the complexities of those systems to the atomic and subatomic levels, although molecular approaches by now have assumed an enormous influence on most fields of biology.

While the distinctions between disciplines are traditional, they are fast becoming less applicable as science crosses the boundaries that once existed. Are efforts to understand biomolecules, the smallest of biological constructions, a facet of chemistry or biology? Are attempts to understand the environmental effects of greenhouse gases a concern of physical science or of biology? It is becoming increasingly irrelevant whether a particular research topic fits neatly into one discipline or another; in fact, many of the most interesting scientific questions and pressing societal issues will require the collective expertise from multiple fields. These areas of overlap are the focus of this report, where the events being studied cannot unambiguously be described as solely contained within the life or physical sciences.

How, then, to best describe the science of the intersection in terms that may be familiar to those working within the constituent fields? Physical scientists might describe it as a composite, a combination of materials (in this case, concepts, tools, and worldviews) with significantly different properties that, when combined, produce something neither could provide separately. Life scientists might describe it as a hybrid, an attempt to produce something new and different through the cross-breeding of ideas and techniques. All would say these intersectional areas of research require expertise and training outside the traditional scope of their disciplines, resulting in new ways of addressing existing problems or new approaches to emerging topics of study. Along with the novelty, though, comes the possibility of frustrations from falling outside the norm of either canonical discipline, such as difficulties in obtaining funding, finding an academic home, or earning tenure.

This report explores both the promises and obstacles associated with research at the intersection of the life and physical sciences. Chapters 2 and 3 examine, in broad terms, the potential opportunities arising from such research for both scientific communities and society in general. Some of the most promising scientific gains at this intersection are explored in Chapter 2, in the form of five Grand

Introduction

Challenges—areas that have the potential to transform our understanding of the natural world. The potential societal benefits that will occur from progress in this area are discussed in Chapter 3. Research at this intersection can help address some of our most urgent societal challenges, from improved sources of food to creative, alternative sources of energy and from improved medical diagnostics and treatments to new, biologically inspired devices that identify and combat biological threats or help to mitigate the effects of climate change.

The next two chapters of this report delve more deeply into research efforts—both now and in the near future—in this intersection of the life and physical sciences. Some intersectional work involves scientists applying concepts developed in one area to issues arising in another. Chapter 4 provides three examples of such crosscutting themes: interactions, dynamics, and pattern formation. Other intersectional efforts involve using tools and techniques originally developed in one arena—principally in the physical sciences—to answer questions in the other. Chapter 5 discusses some of those tools and techniques, including some of the technological advances that will be needed shortly to further research in this area.

Finally, Chapter 6 discusses some of the obstacles that prevent research communities from taking full advantage of opportunities afforded by research at this intersection and proposes a number of recommendations for policy makers, academic institutions, scientists, and others to help reduce those obstacles. Highlighted are new mechanisms for education and training, new models for supporting scientific research, and new means for enhancing coordination between federal agencies.

2

Grand Challenges

As discussed in Chapter 1, this intersectional work between the biological and physical sciences can be characterized as efforts to tackle new issues, typically biological in nature, by adapting ways of addressing problems whose genesis are in another field, typically one of the physical sciences. These adapted ways might be how to conceptualize the problem, or how to evaluate or otherwise draw information out of data, or how to collect the necessary data. In this chapter, the committee sets out five areas of potentially transformative research it believes are particularly susceptible to significant advancement by taking this approach.

Each of these areas, presented in the form of a grand challenge, describes research questions where the scientific challenges are compelling to those in the constituent disciplines, where researchers are poised to make a breakthrough—the goals are attainable in the foreseeable future—and where the payoff of success would be substantial. These are questions whose answers not only will transform our knowledge of the physical world, but also will substantially impact our society.

The committee does not claim that these five grand challenges are the only areas for investment, as there are many areas that could benefit from collaborative attention from the physical and life sciences. But the committee has identified these as among the most urgent, the most important, and the most achievable.

GRAND CHALLENGE 1. SYNTHESIZING LIFELIKE SYSTEMS

Living systems provide proof-of-concept for what can be achieved physically. Can the combined skills and knowledge sets of biological and physical scientists provide greater insight into identifying those structures, capabilities, and processes that form the basis for living systems and then, with that insight, construct systems with some of the characteristics of life that are capable, for example, of synthesizing materials or carrying out functions as yet unseen in natural biology?

For centuries, humans have analyzed the properties of living organisms and built structures to mimic their functions. While most efforts have been at somewhat rudimentary levels, advances in the physical and life sciences now provide us with the technical and scientific sophistication to pursue almost limitless possibilities in this arena. Concepts such as emergent properties, familiar to condensed matter theorists, are helping to describe how biologically complex systems arise from prebiotic chemistry and geochemistry. Other ideas from the physical sciences, such as dynamical systems theory, energy landscapes, and multistability, are helping to explain fundamental issues such as how organisms behave in response to their environments and how information is used to sustain life.[1] Using the knowledge gained in these and other studies, we face the ambitious possibility of generating synthetic units with basic attributes of living matter such as compartmentalization, metabolism, homeostasis, replication, and the capacity for Darwinian evolution. Such self-replicating, evolving organisms have the potential to create more efficient functions for a broad range of applications. At the same time, pursuing this challenge will provide us the opportunity to explore and expand our understanding of the principles of self-replication and evolution.

Any such efforts will require the duplication of essential components of living systems. For example, Darwinian evolution requires a molecular basis for heritable variation, suggesting that any such system must contain a polymer like RNA or DNA with the ability to store and encode information in a simple way. This genetic material must be able to replicate, which requires either a simple autocatalytic system or chemistry that enables spontaneous replication. The scientific community has made some progress toward these goals by, for example, chemically synthesizing natural genomes and then replacing the original genomes in living cells with these synthesized genomes and, in a more bottom-up approach, developing and studying "protocells" as a demonstration of how simple nucleic acids self-replicate within a lipid envelope.

In addition, the products of replication must be held in proximity for some

---

[1] Many of these topics are touched on in more detail in Chapter 4. Also, see National Research Council, *The Role of Theory in Advancing 21st Century Biology*, Washington, D.C.: The National Academies Press, 2008, for related discussions.

time, so that advantageous mutations can exhibit their phenotypes and result in enhanced fitness, leading to differential reproduction and changes in population abundance. Such an achievement will probably require some form of compartmentalization of the components. Cell membranes are possible candidates, since they are composed of a wide variety of amphiphilic molecules. Artificial membranes have been constructed of a wide range of nonbiological materials. However, the specific control of the shape and the osmotic properties when the number of components in the surface and interior of the membrane increases, is unknown. The questions become more interesting as the chemical components diverge more from standard biological components. The section in Chapter 5 entitled "Interactions within Cells" discusses some of the strategies under way to explore how to synthesize such structures.

There is no reason, in principle, why self-reproducing, evolving systems cannot be generated in a wide range of chemical formats. Unfortunately, very little research has systematically approached the chemistry of self-replication based on nonbiological materials. Moreover, a deep understanding of how to efficiently encode and transfer information in highly fluctuating nonequilibrium environments is required. However, attempting to create autonomous synthetic devices capable of self-replication and evolution undoubtedly will generate new principles and tools for synthesizing, assembling, and programming dynamic entities currently unimagined.

GRAND CHALLENGE 2. UNDERSTANDING THE BRAIN

The human brain may be nature's most complex system. Unraveling the mysteries of how it works is one of the greatest of challenges facing the scientific world, and the tools and ideas developed in the physical sciences will play a pivotal role in this undertaking.

One promising approach to understanding the brain is the reverse engineering of neural circuits. This reverse engineering has been accomplished for simple model nervous systems typically consisting of a few dozen cells. For example, neurons in the stomatogastric ganglion of the crab control the musculature of the crab's stomach. Understanding the mechanism of this simple system was accomplished in five steps: (1) cataloguing the different cell and synapse types; (2) measuring their properties; (3) mapping the wiring diagram (the detailed connectivity between neurons); (4) measuring the electrical dynamics of many neurons simultaneously; and (5) creating a model that predicts and simulates behavior.

Understanding the much more complex mammalian brain will require a similar program of reverse engineering albeit on a much larger and more complex scale. Some first progress has recently been made in the form of large-scale, physiologically realistic models of a cortical hypercolumn, of the hippocampal dentate gyrus, and of the entire human brain. Unfortunately, current experimental tools are woefully inadequate for this task, although there are possibilities on the horizon. The physical sciences are particularly adept at developing tools to meet some of the most significant needs—namely, new methods of high-resolution, high-throughput microscopy and imaging to monitor the functions of the brain components, ideally in the intact brain. One key challenge is to trace the thinnest neuronal wires (100 nm and less) throughout the entire brain (tens of millimeters and more in length). This would allow the reconstruction of neuronal shape and the mapping of wiring diagrams (Steps 1-3, above). One such technique, diffusion tensor imaging, is discussed in Chapter 5, in Figure 5-5. Finally, data and understanding are necessary to create a predictive model that will represent a new level of understanding neural function.

GRAND CHALLENGE 3. PREDICTING INDIVIDUAL ORGANISMS' CHARACTERISTICS FROM THEIR DNA SEQUENCE

Individuals belonging to the same species exhibit remarkable diversity in form and function. For example, humans not only look much different from one another, they also differ significantly in their susceptibility to disease. Geographically isolated populations of butterflies develop striking differences in coloration. How much of this variation results from differences in genome sequence, and how much is due to gene-environment interactions? Likewise, how does DNA sequence change in response to interactions with other living things and with the environment? Life and physical scientists will need to work together to develop the theory and modeling to understand these phenomena.

Ultimately, the blueprint for form and function lies in an organism's DNA sequence. A major challenge is to understand the relationship between the DNA sequence (genotype) and the individual's characteristics (phenotype). Small differences in genotype can interact to produce large changes in phenotype. To understand how genetics underlies phenotype, we need quantitative models of genetic interactions.

DNA sequences change over time in response to selective pressures. Selection acts at the level of the individual but is defined by the individual's fitness relative to the rest of the population, which, in turn, is affected by the results of selection. Thus, the outcome of selection manifests itself at the level of the population. The population exists in a particular environmental niche, and changes in the environment can affect the population; likewise, the population can have dramatic affects on the surrounding environment. All of these components feed back to affect the survival and therefore selection of the individual (and thus its genes). Interactions between the environment and genes, interactions between organisms of the same and different species, interactions between different species and their environment, and how all of these interactions iteratively feed back to alter the environment and thus selection must be understood. Here again, interactions between life scientists and physical scientists will be needed to develop models that help to understand these phenomena.

Thanks to high-throughput DNA sequencing, we now possess the complete code of hundreds of organisms, including humans. A major challenge going forward is to decipher the principles underlying the organization of the information in the DNA. Interactions between DNA elements introduce a combinatorial wealth of possibilities for stringing sequences together to impart complex patterns of gene expression. Deciphering the logic of gene regulation will require theorists, at times drawing on theoretical constructs originally developed in the physical sciences, to work hand in hand with molecular biologists. The gains from those efforts will allow us to more fully exploit the information we have garnered from the human genome sequence and learn how it relates to health and disease, aging, and the quality of life.

GRAND CHALLENGE 4. INTERACTIONS OF THE EARTH, ITS CLIMATE, AND THE BIOSPHERE

Many of the most challenging and potentially most important questions in science involve interactions among actors or processes governed by strikingly different mechanisms. Often these processes operate on much different scales of time and space, unfolding over spatial scales ranging from the microscopic to the global and on temporal scales from fractions of a second to many millions of years. In the past, these questions were often discussed but rarely tackled in a comprehensive way. The difficulties in undertaking quantitative studies of these matters typically loomed so large that the interactions were simply considered to be outside the boundaries of the inquiry at hand. Questions about, for example, the role of microorganisms in shaping Earth's near-surface environment or the role of forests in modulating the tempo of glacial-interglacial transitions were, until recently, too complex, too multidisciplinary, and too multiscale for profound treatment with available scientific tools. Now, however, they are ripe for joint investigation by life and physical scientists.

Such joint efforts are needed, in part because many of the processes involve intricate, continual interactions between physical and biological parts of the system. Consider, for example, the global nitrogen cycle. At the spatial scale of microns, the transformation of organic nitrogen to NH_4^+, NO_3^-, N_2, NO, and N_2O appears to be controlled by, among other things, the microscale soil moisture and the respiration rate of the organisms in the microsite. But these transformations interact to adjust soil fertility over thousands of years. The compounds released have the potential to influence global climate and vast stretches of aquatic habitat. As with most of the interactions between the living and the nonliving parts of a system, even a complete knowledge of all the separate parts still falls short of allowing us to understand how they interact over different timescales. The challenges of understanding the interactions among living and nonliving parts of the Earth system are often exacerbated by the consequences of human actions. In the example of the global nitrogen cycle, human actions have more than doubled the amount of biologically available nitrogen moving through the Earth system, making it difficult to observe or understand the system free of anthropogenic influences.

Many of the foundations for rapid progress in addressing these questions are either in place or nearly in place. Increasingly powerful computer models can link, for example, the global climate with site-to-site variations in the biological parts of the global carbon cycle, which, in turn, feed back to the global climate. Space-based sensors provide access to a growing set of observations at regional to global scales, facilitating the coordinated analysis of, for example, shifts in the locations of the major biomes, changes in the distribution of pests and pathogens, changes in regional water balance, and feedbacks to climate change. See, for example, Figure 5-6 in Chapter 5, on satellite imaging technologies. Other approaches to observing the physical parts of the system—for example, from familiar computerized axial tomography (CAT) scans to the less-well-known spectroscopies such as extended X-ray absorption fine structure (EXAFS), X-ray absorption near edge structure (XANES), and secondary ion mass spectroscopy (SIMS)—provide access to sites that are smaller, bigger, more protected, or more complex than have been accessible so far. The range of tools either available or becoming available in molecular biology is increasing the power and sweep of experimental studies by many orders of magnitude. In short, the scientific community has the questions, the tools, and the concepts to effectively tackle the questions surrounding interactions between living and nonliving parts of the Earth system.

Broad question areas related to this grand challenge that are ripe for future breakthroughs include the following:

- How have life and Earth coevolved over time? For instance, what climate changes can be attributed to the evolution of key biochemical processes, and what events in evolutionary biology were triggered by geochemical events?
- How are changes in Earth's climate affecting terrestrial and marine biology? How will changes to these ecosystems feed back into the climate system?
- How does the physical diversity of Earth's habitats help to control the biological diversity of its organisms, and how does biological diversity alter the functioning of ecosystems?
- How does the ability of organisms and ecosystems to cope with change depend on the complexity of the physical environment?

- Through what mechanisms are interactions among the living and nonliving parts of the Earth system coordinated, and when does that coordination fundamentally shift or disappear?
- In an era of increasingly pervasive human influence on physical and biological components of the Earth system, what are the most effective strategies for maintaining the integrity of natural systems and the services they provide?

GRAND CHALLENGE 5. UNDERSTANDING BIOLOGICAL DIVERSITY

Life on Earth is astoundingly diverse, a result of evolution. While such diversity has been recognized for centuries, its role in the health of a species or ecosystem has only recently begun to be studied. An understanding of diversity is becoming increasingly important as human activity has a larger and larger impact on the natural world. The modeling capabilities and tools of the physical sciences will play a critical role in such studies.

Diversity appears in the natural world at many levels. A single multicellular organism can consist of more than 10^{15} cells, divided into many different organs. In some organs, no two cells are identical. Within a single species, individual organisms vary extensively at the DNA sequence level, and this translates into substantial diversity in appearance and behavior. Dogs are in breeds that range from Great Danes to Chihuahuas. However, we have only rough estimates of the extent of existing diversity. While the diversity of well-studied groups like birds and mammals is generally well known, data on the diversity of insects, microorganisms, and marine invertebrates are thin. Recent estimates conclude that only 5-10 percent of such species have been classified.[2] Diversity within populations and organisms is known for very few taxa.

Within a species, diversity in the sequence of genes protects against extinction by infectious agents or predators and may allow species to function efficiently across a wider range of environmental conditions. The role of diversity in the functioning of ecosystems is only beginning to be understood. A number of studies indicate that plant communities tend to be more productive, more resistant to biological invasives, or less sensitive to disturbance when they are more diverse. Some evidence supports the hypothesis that increasing diversity increases the probability that a community contains at least one well-adapted species (a sampling effect). Other evidence points to a complementarity in which diversity allows species to forage more efficiently for resources. For example, a more diverse community could contain species active at different times of the year, species that extract water and

---

[2] Available at http://www.globalchange.umich.edu/globalchange2/current/lectures/biodiversity/biodiversity.html#readings. Accessed April 21, 2009.

nutrients from different depths in the soil, or microbes that specialize on different substrates.

It is likely that diversity influences ecosystem function through a wide range of mechanisms. Some are structural (rooting depth), some behavioral (feeding time), and some biochemical (optimum temperature for an enzyme).

Human activities have tended to reduce the diversity of organisms and ecosystems, in some cases dramatically. The widespread use of limited numbers of cultivars has rendered many of our agricultural crops potentially very sensitive to disease outbreaks or climate fluctuations. The same concern exists for commercial livestock. Now, in an era of rapid global changes driven by human actions, the role of diversity and the processes that maintain diversity take on dramatic new importance. Many kinds of human impacts on the natural world, especially landscape fragmentation, climate change, pollution of air, soil, and water, and stimulation of biological invasives, threaten to decrease biological diversity. At the same time, the novel habitats these impacts create may need the maximum possible biodiversity if they are to cope effectively in the novel conditions. The Intergovernmental Panel on Climate Change (IPCC) estimates that a global average warming of as little as 2°C could commit 30 percent of the world's species to extinction. We do not know the mechanism, the timetable, or the consequences, nor do we know the features that make some communities extremely resilient, while others collapse with only modest forcing.

Our ability to sequence DNA rapidly and inexpensively is increasing exponentially. This promises to provide the capability to extensively characterize the diversity of species and to allow predicting its functional consequences. We have begun to be able to determine the DNA sequences of complete simple ecosystems, and this will inevitably progress to more complete and interesting communities and organisms. These tools, combined with advanced techniques for analyzing the physical and modeling tools of ecology, offer the potential for huge breakthroughs in coming decades. A clearer picture of the role of diversity in ecosystem functioning will make it feasible to specify the diversity level required to secure sustainable provision of key services. A clearer understanding of existing diversity and the factors that lead to extinction should enable a suite of strategies for protecting diversity in key areas.

For this challenge, there are opportunities for the biological and physical sciences to interact at a range of levels. Physics will be able to extend the tools needed to quantify diversity quickly and cheaply, which is crucial for understanding the consequences of altering diversity for ecosystem functioning.

3

Societal Challenges

Earth has been called an engineered planet. It is not a stretch to say that the long-term well-being of nearly 7 billion people is unimaginable without engineering and modern technology. Humans are dependent on a complex mix of natural resources and manufacturing processes combined with technological infrastructure for food, water, shelter, and other needs. But this infrastructure and those processes are strained by global development, population growth, climate change, and competition for limited resources. Sustaining the integrity and functionality of Earth's natural systems while maintaining and even enhancing the well-being of humans will require both new technologies and a much deeper understanding of the Earth's natural processes. The scientific and technological understandings that come from research at the intersection of the physical and life sciences have the potential of helping to meet those needs.

This chapter discusses a few such applications in agriculture, energy, climate, biomedicine, and novel materials. While these examples may not be comprehensive, they illustrate some of the many ways in which research at the intersection of the physical and life sciences has addressed and will continue to address some of our most difficult societal challenges.[1] The committee is mindful of the fact that societal benefits will emerge not only from advances in scientific research at this life sciences/physical sciences intersection but also from increased interactions between the life sciences and various engineering disciplines. However, as important as the

---

[1] Many of the applications discussed in this chapter are also highlighted in the recent NRC report *A New Biology for the 21st Century* (National Research Council, 2009).

latter set of interactions might be for society, they are beyond the scope of this report.[2]

IDENTIFYING AND COMBATING BIOLOGICAL THREATS

In the past few years many people have come to think of biological threats only in the context of war or terrorism. However, natural biological threats have existed throughout history and still exist today. We need to look no further than recent outbreaks of sudden acute respiratory syndrome (SARS), swine influenza, and avian influenza to see that disease organisms can evolve, adapt, and cause epidemics.

Early Detection and Intervention

The appropriate defense always involves a medicinal attack on such organisms; however, early response to a disease depends on early recognition. In some cases, this means recognition of a weak disease "signal" in the noise of everyday life. To that end, information scientists have begun to collect data about symptoms posted on public health and popular medicine Web sites and have used that information to look for increased incidences and the appearance of clustered events.

In addition to its use in detecting chemical threats from a distance, spectroscopy can be used for remote sensing of pollutant molecules from individual automobiles under normal driving conditions to find and eliminate the worst polluters. Similarly, the remote, noninvasive identification of diseased individuals, by detecting thermal or chemical signals, is used to find and give early treatment to people who are not otherwise recognizably sick. Today there are devices that can "smell" cancer by detecting the metabolites of specific cancer cells (Rovner, 2008). Such identification and treatment could reduce the severity of disease in an individual or mitigate the spread of infectious disease in the larger population. This becomes increasingly important as people fly around the globe within a day and geographical barriers to the spread of disease fall.

In addition to the direct effects of disease in individuals, humans' experience with "mad cow" disease and the swine and avian influenza viruses points to the importance of animal health and welfare and highlights the threat that zoonotic diseases such as the West Nile virus pose to the larger human population. This threat can be to health itself or to the economy. In 2001, Great Britain fought a battle against foot and mouth disease (FMD), a highly contagious virus, slaughter-

---

[2] Some of these areas were explored in the recently released NRC report *Inspired by Biology—From Molecules to Materials to Machines* (National Research Council, 2008).

ing over 4 million animals to control the disease.[3] Prevention and suppression of FMD and other, similarly infectious pathogens depends on early detection, quarantine, and/or destruction of disease carriers. The rapid remote sensing of such animals could help prevent or reduce the negative impact of these diseases.

Using techniques related to speech recognition, mathematicians have worked to map the genetic similarity of influenza viruses. Using color and spatial distribution, subtle evolved differences in virus strains can be identified, and decisions about the potential effectiveness of vaccines become easier (Enserink, 2008). Obviously both symptom recognition and pathogen strain identification are also very useful in the defense against any man-made biological.

Prediction of Susceptibility to Disease and Its Prevention

Historically, vaccination, simple sanitation and the availability of clean drinking water were perhaps the most important developments in disease prevention. However, these methods are useful only for preventing infectious disease, and we are discovering their limitations and drawbacks. Prevention of disease, especially noninfectious disease, remains a critical issue.

The impact of fetal nutrition on adult susceptibility to obesity points to the potential benefit of better understanding the sources and causes of noninfectious disease. As these causes are found, we can more easily identify and treat individuals most likely to develop a disease. Doing so entails both a deeper understanding of the biology of disease inception and the physical science of detecting susceptibility.

Beyond detection of susceptibility, however, is the need to understand how conclusions drawn from populations translate to therapy for individuals. Here is the greatest need for personalized medicine—that is, diagnosis and treatment tailored to a person's genetic makeup and potentially to environmental differences as well. Personalized medicine in simple forms is already used—if a breast cancer patient's tumor is estrogen-receptor positive, then the patient may benefit from treatment with tamoxifen, which prevents estrogen from binding to the receptor and stimulating growth.

In the future, analysis of an individual's gene expression patterns using microarrays or even sequencing of his or her entire genome, which will require enormous advances in chemistry and engineering, should allow physicians to diagnose and treat disease with greater accuracy. A greater understanding of the DNA sequence changes that impact protein-coding or RNA-coding should also allow these processes to be mathematically modeled.

---

[3] Information from "Farm incomes in the United Kingdom 2001/2002." Available at https://statistics.defra.gov.uk/esg/publications/fiuk/2002/FIUK_complete.pdf. Accessed January 28, 2009.

CLIMATE AND ITS INTERFACE WITH BIOLOGY

Energy generation, energy use, the environment, and Earth's climate are inextricably bound, and the consequences of these interactions are broad, ranging from human comfort and disease to wildlife and agriculture (IPCC, 2007a). While climate science has reached a level of sophistication sufficient to document components of these interactions, such as how humans impact climate (IPCC, 2007b) and how projects impact ecosystems, it is not yet capable of understanding the full consequences of the actions we might take to mitigate those impacts while simultaneously maintaining our energy supply.

The technical challenges associated with energy and climate policy require a deeper understanding of the intersection between physical and biological systems and, therefore, the physical and biological sciences. This would be particularly true for global engineering such as modifying the atmosphere to be more reflective or seeding the oceans with iron to induce algae growth and carbon dioxide uptake. Any such efforts must take into account both direct and indirect aspects of the complex interactions that link the climate, ecosystems, and the oceans (Field et al., 2007).

Complex Feedback Loops in Climate Science

Some of the most pressing scientific questions about climate change concern the risk that warming may lead to large releases of carbon as CO_2 or methane from land and ocean stocks, reinforcing the warming trend (Gruber et al., 2004). Such releases and the consequent, possible shift in the relative mix of carbon dioxide and other greenhouse gases increase the possibility of further feedback. Finally, changes in vegetation cover caused by warming will affect absorption rates of solar radiation at the Earth's surface thereby creating another class of feedbacks that may play a large role in amplifying or perhaps suppressing climate change (Gibbard et al., 2005).

The scientific and societal challenge of understanding the role of physical/biological interactions in climate change is at least as profound in the area of solutions as it is in impacts. Essentially all of the possible approaches for offsetting emissions of greenhouse gases, including biological sequestration, decreased deforestation, and geological and deep ocean sequestration, involve changes to biological systems, with the possibility of indirect impacts that either feed back to climate change or alter the delivery of ecosystem goods and services.

Implications of Renewable Energy

Even energy technologies that are not explicitly based on combustion or biology will likely have important impacts at the intersection of physical and biological sciences. For example, large-scale harvesting of wind energy may alter atmospheric transport and turbulence. Large-scale solar collection will alter energy absorption at Earth's surface, the partitioning of energy into evaporation and sensible heat, and, at least in some cases, the light available for photosynthesis. Large-scale hydro, wave, or tide power will likely have widespread effects on hydrology and on the organisms and people that use the water resources. Thoroughly assessing the impacts of these technologies to ensure that they create the smallest possible set of environmental problems is a key challenge for the future, one that is potentially as important as understanding the mechanisms and impacts of climate change.

MEDICINE

The importance of physics and chemistry to medicine is well known. The discovery of X rays and nuclear magnetic resonance by physicists in the first half of the twentieth century led the way to the diagnostic X rays, CAT scans, and MRIs of today. The ability of chemists to isolate, analyze, and synthesize complex organic molecules led to the modern pharmaceutical industry. Not only have these historical trends continued, they have accelerated, and many new medicines have come from advances in our understanding of molecular interactions, chemical reactivity, and synthesis. The future will require further development of physics techniques and new understandings in chemistry and chemical methods to enhance the efficiency of industrial syntheses and reduce the generation of by-products that harm the environment.

Imaging

Physicists and applied physicists are working out the theory and design for improved imaging of biologically important entities, from the human body as a whole to individual cells and molecules. Applications range from diagnostics to minimally invasive surgery or radiation treatment.

Advances in high-resolution light microscopy allow researchers to image the position and movement of molecules at the nanometer scale in real time (Pinaud and Dehan, 2008). Biological processes involve the binding of molecules, large or small, and sometimes a number of them simultaneously, to large, protein-based receptors. Using novel imaging techniques, molecules may be found to exist in close proximity with one another in a cell. This could mean that the molecules act together or on similar structures in that cell and could aid in understanding the binding mechanisms.

When the equations for x-ray diffraction were first elucidated, the work seemed like pure, esoteric physics. Now biophysicists at the national laboratories

(Brookhaven, Argonne, and Lawrence Berkeley) and elsewhere routinely use high-powered synchrotron beamlines to see individual atoms of drug molecules binding to their protein targets. This allows chemists to modify the structures of the drugs based on images and not on live patient response, speeding up the process of making drugs more specific and more powerful, and giving hope for a new constellation of antibiotics aimed at drug-resistant pathogens.

In some cases, common and necessary imaging techniques are not without the potential to do harm. For well over 100 years, X rays have been used for diagnostic and therapeutic processes, sometimes damaging vital organs. The mechanisms of radiation-induced damage involve cells committing suicide via a process called apoptosis. Recently, drugs have been developed that can trigger natural cellular mechanisms that resist this process and thereby mitigate such damage (Bhattacharjee, 2008).

Treatment and Devices

Research at the intersection of the physical and life sciences has made significant contributions to the development of new treatments and devices, a few examples of which are briefly discussed here. For diabetics, advances in nanotechnology have resulted in nano-scaled dispensers of insulin that, when combined with continuous monitoring of blood sugar levels, allow for the administering of the right amount of insulin in a continuous manner. Further, joint efforts by chemists and microbiologists are seeking to understand the processes by which insulin is generated and other hormonal activity is regulated by the pancreas, with the ultimate goal of creating artificial pancreas. Such efforts offer hope to diabetics whose disease has caused the destruction of that organ (Halford, 2008).

Intersectional research also has shed light on the interaction between mind and machine. As an example, the implantation of a small electrical interface into a monkey's brain allows the monkey to control a prosthetic arm by its thoughts. A computer analyzes the response of the monkey's brain to a stimulus and transmits an electrical impulse to the prosthesis. With practice, a monkey learns what kind of response is needed to operate the arm (Cary, 2008).

AGRICULTURE AS A RESOURCE FOR FOOD AND ENERGY

Since the first use of fire, fuel has consisted of biological products derived from plants, which are in turn created via photosynthesis. Fuels such as wood, crude oil, or coal are not primary fuels; rather they are batteries, storing energy from the Sun.

Historically, it has always been sufficient for humans to harvest these biobased materials, whether they are found above or beneath Earth's crust, and simply burn them. In the twenty-first century, however, demand for energy will likely outstrip the availability of these sources, and their unfettered use will be complicated by concerns about products of combustion (see discussion of climate, below).

Researchers and technologists have developed non-bio-based sources of energy and means to store it. However in many cases these technologies utilize toxic or less abundant substances—for example platinum or palladium catalysts, gallium arsenide solar collectors, and nickel, cadmium, lithium, or even tried-and-true lead batteries—whose supply or disposal is problematic.

Increasingly, the fuels needed to meet energy demand are being farmed—whether plant oils for biodiesel or plant sugars for fermentation into ethanol. But dependence on agriculture for fuel carries hidden financial, environmental, and security costs. For one thing, a large percentage of the arable land in the world is already in farm production. Additionally, fresh water needed for irrigation is unevenly distributed and in many places in short supply. And, as agriculture becomes the source of fuel as well as food, the supply of both becomes more vulnerable to common weather, biological, and environmental risks. In short, agriculture as it is currently practiced can reliably offset only a small fraction of the global growing demand for energy (Field et al., 2008).

Current research is focused on understanding the biological mechanisms that generate usable fuel so that more fuel material can be produced from incident sunlight. This would mean growing better plants, especially those that will not be used for food, or by adapting their photosynthetic infrastructure to the manufacture of fuel.

Building Better Plants and Getting More Out of Them

Currently agriculture-based liquid fuels are created by fermenting plant sugars—usually corn or sugarcane—into ethanol or reforming plant oils into biodiesel. But, short of burning plant waste for process energy, as is done with sugarcane bagasse, both processes utilize fruit or seeds that constitute less than half the mass of the plant while ignoring the greater mass of stalks and leaves. Other research has led to biobased production of more energy-dense fuels than ethanol (BP, 2008). Enzymatic depolymerization of the cellulose and hemicellulose of those stalks and leaves, yielding fermentable sugars, is becoming more cost effective, as is thermal decomposition of the same materials to yield raw materials for industrial fuel synthesis. These near-term opportunities can be paired with longer-term strategies for harnessing the potential of plants.

Identification, isolation, and manufacture of the key enzymes that catalyze depolymerization of cellulose lie at the intersection of the biological, physical, and engineering sciences, as do efforts to better understand natural systems and how to modify them to yield materials of greater catalytic activity. Low-resource perennial plants such as grasses or algae that grow quickly can be harvested to produce relatively large amounts of plant mass to feed these cellulose depolymerization processes. Better understanding of microorganism culture, the productive potential of soils, and the locations where they provide the greatest net benefit are keys to better utilization of these nontraditional crops.

Researchers are also beginning to isolate "biofuel" genes—that is, genes responsible for increasing plant production of sugars, cellulose, and oil—as well as more efficient enzymes to process cellulose, and to grow more drought- and salt-resistant plants and to increase planting density (Kintisch, 2008). Understanding how these genes work has the potential to increase our understanding of how physical scientists and engineers might utilize similar processes for artificial systems. Moreover, because of the resilience of agricultural pests and their evolutionary potential, it will be necessary to understand more completely the ways plants protect themselves and how we can help protect them in the face of a fragile and hostile environment.

Hydrogenases and Synthetic Photosynthesis

Enzymes and organisms that could produce hydrogen in biological systems have been known for 75 years. These enzymes, which allow the splitting of water into hydrogen and oxygen, are complex and only now becoming better understood. In their ability to use the energy from light to power this reaction, these enzymes are analogous to the complex plant structures responsible for photosynthesis.

Of the opportunities at the intersection of the physical and biological sciences, perhaps none has such intriguing potential as understanding, controlling, and improving photosynthesis, with the goal of decoupling it from plants. While photosynthesis is so common as to be all around us, understanding it and adapting it remains a huge challenge.

Effectively, plants use catalysts derived from common metals to harvest sunlight, to split water into oxygen and hydrogen, and to generate plant material from carbon dioxide. And while the photochemical yield in photosynthesis is high, the steps that actually fix CO_2 and lead to the production of plant material—the biomass that will become fuel—are much less efficient.

One research goal is to isolate the active species responsible for the various reactions of photosynthesis, conveniently split water, separate the resulting oxygen and hydrogen or other reduced species, then use those reduced species as a fuel in themselves or as the raw material in a fuel-making process. Other goals include enhancement of the local flora to increase the CO_2-fixing efficiency.

Industrially usable catalysts inspired by this chemistry with significantly greater yield than natural systems could produce fuel from waste carbon dioxide as plants do. Such technology would simultaneously address a portion of the world's growing demand for energy and rid the planet of atmospheric carbon dioxide.

Beyond Combustion

Much of the fuel we burn is used to generate electricity, typically at about 30 percent efficiency. However, combustion has environmental drawbacks beyond the emission of carbon dioxide. Depending on conditions, carcinogenic and polluting products of incomplete combustion, such as particulates, nitrogen oxides, and polycyclic aromatic hydrocarbons, can also be produced.

Biological systems generate their energy directly in mitochondria at efficiencies of near 90 percent. The energy chemistry in mitochondria is analogous to that in fuel cells, wherein hydrogen and oxygen in the presence of catalysts are converted to water while generating electricity.

Understanding the common-metal catalysts in mitochondria and their adaptation to fuel cells, especially for mobile applications, could simultaneously reduce energy use by improving efficiency and reduce their unwanted by-products of combustion. There are encouraging results in this field as well (Winther-Jensen et al., 2008).

MATERIALS SCIENCE

The commercial development and production of materials mimicking biological systems has been the focus of much industrial research effort. However, many biological materials remain outside commercial reproduction capability. For instance, long spider silk proteins, as fabricated into strands, have the tensile strength of steel, yet the structure of spider silk. As desirable as these characteristics are, the commercial manufacture of spider silk as an advanced material continues to elude engineers. Composite materials—steel-reinforced concrete or glass-fiber-reinforced plastic—have been staples of construction and engineering for years; yet they do not achieve the strength and toughness of biocomposites such as bone or tooth enamel.

Progress is being made in understanding the structure, physical properties, and means by which these materials are fabricated. Synthetic biology has been used to increase production under controlled conditions (UCSF, 2008). Further efforts in replication, manufacture, or modification could lead to lighter, stronger, more resilient, and, moreover, biodegradable engineered materials. This intriguing area is the subject of a recent NRC report on biomimetic materials (NRC, 2008).

OPPORTUNITIES

As modern science dawned hundreds of years ago, its practitioners hoped to understand the large questions of life, including its origin and the maintenance of youth and health. Over time, scientists found those questions too complex to answer with the information that was available, and broke the sciences into smaller disciplines in the physical and biological realms.

Today, after 200 years of studying those more specific disciplines, many of the most fascinating and important problems lie at the intersection and reintegration of these two realms. And, for the first time, twenty-first-century scientists have the tools and knowledge base to address the large issues that impact the maintenance and quality of life. While this will not be easy, the significant advances in agriculture, energy, medicine, and understanding of climate that appear to be increasingly likely by such intersectional work are required to sustain not only our way of life but also its very existence.

REFERENCES

Bhattacharjee, Y., 2008. Drug bestows radiation resistance on mice and monkeys, *Science* 320: 163.
BP, 2008. http://www.bp.com/liveassets/bp_internet/globalbp/STAGING/global_assets/downloads/B/Bio_biobutanol_fact_sheet_jun06.pdf. Accessed August 21, 2008.
Cary, B., 2008. Monkeys control robot arm with their thoughts, *New York Times*, May 29, 2008.
Enserink, M., 2008. Mapmaker for the world of influenza. *Science* 320: 310.
Field, C.B., J.E. Campbell, and D.B. Lobell, 2008. Biomass energy: the scale of the potential resource. *Trends in Ecology & Evolution* 23: 65-72.
Field, C.B., D.B. Lobell, H.A. Peters, and N.R. Chiariello, 2007. Feedbacks of terrestrial ecosystems to climate change. *Annual Review of Environment and Resources* 32: 1-29.
Gibbard, S., K. Caldeira, G. Bala, T.J. Phillips, and M. Wickett, 2005. Climate effects of global land cover change, *Geophysical Research Letters* 32: L23705.
Gruber, N., P. Friedlingstein, C.B. Field, R. Valentini, M. Heimann, J.E. Richey, P. Romero-Lankao, E.D. Schulze, and C.T.A. Chen, 2004. The vulnerability of the carbon cycle in the 21st century: An assessment of carbon-climate-human interactions. *The Global Carbon Cycle: Integrating Humans, Climate, and the Natural World*: 46-76. Washington, D.C.: Island Press.
Halford, B., 2008. Perfecting an artificial pancreas, *Chemical Engineering News* 86: 46-47.
IPCC, 2007a. *Climate Change 2007: The Physical Science Basis.* Contribution of Working Group I to the Fourth Assessment Report of the Intergovernmental Panel on Climate Change, S. Solomon, D. Qin, M. Manning, Z. Chen, M. Marquis, K. B. Avery, M. Tignor, and H. L. Miller, editors. Cambridge, United Kingdom and New York, NY, USA: Cambridge University Press.
IPCC, 2007b. Summary for Policymakers in *Climate Change 2007: Impacts, Adaptation and Vulnerability*. Contribution of Working Group II to the Fourth Assessment Report of the Intergovernmental Panel on Climate Change, M. L. Parry, O.F. Canziani, J. P. Palutikof, P. J. van der Linden, and C. E. Hanson, editors. Cambridge, UK: Cambridge University Press.
Kintisch, E., 2008. Sowing the seeds for high-energy plants, *Science* 320: 478.
National Research Council, 2009. *A New Biology for the 21st Century*. Washington, D.C.: The National Academies Press.
National Research Council, 2008. *Inspired by Biology: From Molecules to Materials to Machines*. Washington, D.C.: National Academies Press.
Pinaud, F., and M. Dehan, 2008. Zooming into live cells, *Science* 320: 187-188.
Rovner, S.L., 2008. Sniffing out cancer, *Chemical Engineering News* 22**:** 81-83.
UCSF, 2008. http://www.ucsf.edu/science-cafe/conversations/voigt/ Accessed August 21, 2008.
Winther-Jensen, B., O. Winther-Jensen, M. Forsyth, and D.R. MacFarlane, 2008. High rates of oxygen reduction over a vapor phase–polymerized PEDOT electrode, *Science* 321: 671-674.

4

Common Themes at the Intersection of Biological and Physical Sciences

In this chapter the committee explores some of the problems and conceptual challenges at the intersection of the biological and physical sciences. Without attempting an exhaustive survey of the subject, it covers a handful of examples that illustrate both the importance of the open scientific problems and some of the breakthroughs that are occurring as a result of inquiries that bridge this intersection.

INTERACTION AND INFORMATION: FROM MOLECULES TO ORGANISMS AND BEYOND

The modern era of biology began when, together, a biologist and a physicist uncovered the nature of the interactions holding together the strands of DNA in the famous double helix. The relatively simple interactions between different pairs of nucleotides immediately revealed the nearly infinite information storage capacity of the DNA heteropolymer and defined, for the first time, the intimate connection between interaction and information that makes up the fabric of living matter. Today, 60 years later, the challenge in studying living matter is to produce a framework for understanding the highly organized and information-rich biological structures that are engaged not only in the acquisition and conversion of metabolic energy but also in the acquisition and transfer of information. Unraveling the complexity of living systems is a challenge that requires not only the creative application of ideas and tools for interacting systems but also the development of new conceptual and mathematical frameworks that can incorporate information and information fluxes alongside the existing thermodynamic and statistical principles of physical science. It is at this interface of biology with physics and information theory that the fundamental principles governing living matter are likely to be discovered.

Biological complexity is built on specific interactions between molecules, and these interactions are linked to each other and held in balance through complex networks. These networks underpin the regulation and signaling that govern intracellular function and multicellular behavior all the way to the development of the organism, and their multilayered complexity makes studying the systems challenging.

On the smallest level, interactions are mediated by molecular forces (hydrogen bonding, electrostatics, hydrophobicity), which form the physicochemical basis of molecular recognition between polynucleotide and polypeptide structures. Although we understand the basic laws governing these forces, using these laws to reliably predict specific, complex intermolecular interactions and tracking the effect of the intermolecular interaction to the behavior of a whole organism remain a challenge.

To deal with some of these challenges, ab initio approaches are now frequently complemented by data-driven bioinformatics that analyze and compare empirical data to untangle the interactions between numerous related interacting pairs of molecules. This combination of approaches weaves together ideas and methods from computer science, statistics, physics, and biology, and researchers use them to reveal the patterns (called "code" by some) underpinning the interaction (see Figure 4-1).

Bioinformatic studies provide supramolecular-level descriptions in which, instead of the basic interatomic forces, one works with interaction profiles—that is, the strengths of interactions with different possible partners that de facto define biological molecules. Indeed, characterization of such interaction profiles allows an understanding of interactions at the level of the whole organism and bioinformatic approaches are now extensively used for identifying regulatory targets of transcription factors—proteins that control gene expression. This approach is delivering ingenious quantitative tools that enable us to extract enhanced knowledge from existing biological data in new ways.

The story does not end with describing the specific interactions, however. As mentioned above, the interactions between biomolecules are the building blocks of molecular and genetic networks, and the networks must also be studied and understood. For example, a covalent modification of a specific protein via the enzymatic activity of another (e.g., phosphorylation catalyzed by a kinase) might trigger enzymatic activity of the target protein or cause its re-localization within the cell. In genetic networks, regulation of gene expression and protein synthesis are controlled through the action of transcription factors and recently discov-

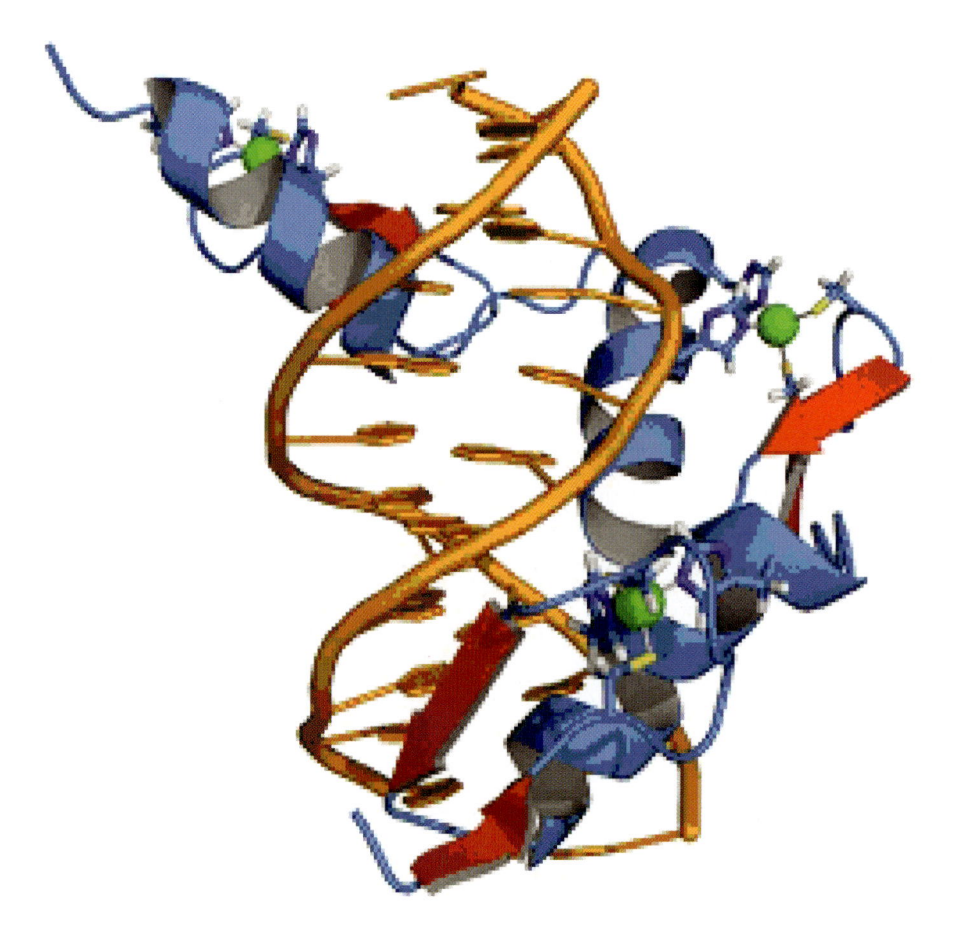

FIGURE 4-1
Zinc Fingers

Zinc finger proteins form a ubiquitous family of transcription factors—proteins that bind DNA in a sequence-specific manner and regulate gene expression. They have a remarkable modularity, which allows mixing and matching of up to six DNA binding domains, fostering highly specific targeting of a diverse set of DNA motifs. Because of this feature, zinc fingers hold great promise as tools for the precise control of gene regulation and have potentially numerous and profoundly important medical applications (Klug, 2005). The design of engineered zinc fingers for practical applications requires a thorough understanding of the interaction "code" that defines transcription factor/DNA binding specificity. Deciphering this code brings together structural analysis with bioinformatics and ab initio computational modeling studies (Paillard et al., 2004). The figure shows the structure of the transcription factor protein Zif268 (blue) containing three zinc fingers in complex with DNA (orange).
SOURCE: Pavletich and Pabo, 1991; reprinted with permission from the American Association for the Advancement of Science.

ered micro-RNAs. These and other methods of regulating protein abundance and activity are the links in the causative chains forming cascades and networks that propagate and modulate the effects of different stimuli.

Networks mediated by a mitogen-activated protein (MAP) kinase are another excellent example of this type of network. This family of signaling proteins controls the regulation of diverse processes, ranging from the expression of genes required for a yeast cell to adapt to the carbon sources in the environment to the transcription of proto-oncogenes in the development of cancer, to programmed cell death. Over a dozen MAP kinase family members have been discovered in mammals alone. Each MAP kinase cascade consists of a minimum of three kinase enzymes that, in response to particular stimuli, are activated in series by the MAP kinase above it in the signaling cascade. MAP kinase pathways transfer information to particular effectors that perform a number of functions, including integrating information channeled from other regulatory pathways into the MAP kinase pathway, amplifying particular signals under particular conditions, and precisely directing an array of discrete response patterns. The inherent complexity of the system challenges our ability to characterize and understand the functions and mechanisms of the individual pathways. MAP kinases are only one example of the interconnectedness of the networks of cellular regulatory pathways, and studying such systems provides interesting challenges for both physical and life scientists, because new quantitative concepts and approaches will be necessary to disentangle these interacting network phenomena.

Interaction networks not only exist within cells but also extend outward to coordinate cellular responses to the environment and to coordinate the behavior of groups of cells through intercellular interactions. Paradigmatic examples include bacterial chemical sensing of the environment (chemotaxis) and cell-to-cell interactions through chemical signaling (quorum sensing), illustrated in Figure 4-2. The two bacterial behaviors are unified by the key role played by information transfer. Chemotaxis refers to the ability of bacteria to swim in a biased direction—either toward a gradient of a nutritious compound or away from a gradient of a noxious one. The chemotaxis sensory circuit relies on receptors mounted on the membrane, which bind to attractant or repellant molecules. These binding events elicit a protein phosphorylation cascade, which causes motor proteins to switch the rotation of the flagellar apparatus such that, depending on which direction it is spinning, the bacterium either tumbles or is propelled forward. Quorum sensing is a process of bacterial cell-to-cell communication that involves production and detection of threshold concentrations of signal molecules called autoinducers. The accumulation of a stimulatory concentration of an extracellular autoinducer occurs only when a sufficient number of cells—a "quorum"—is present. Thus, the process is a mechanism that allows bacteria to collectively regulate gene expression and thereby function as multicellular organisms. In both chemotaxis and quorum-

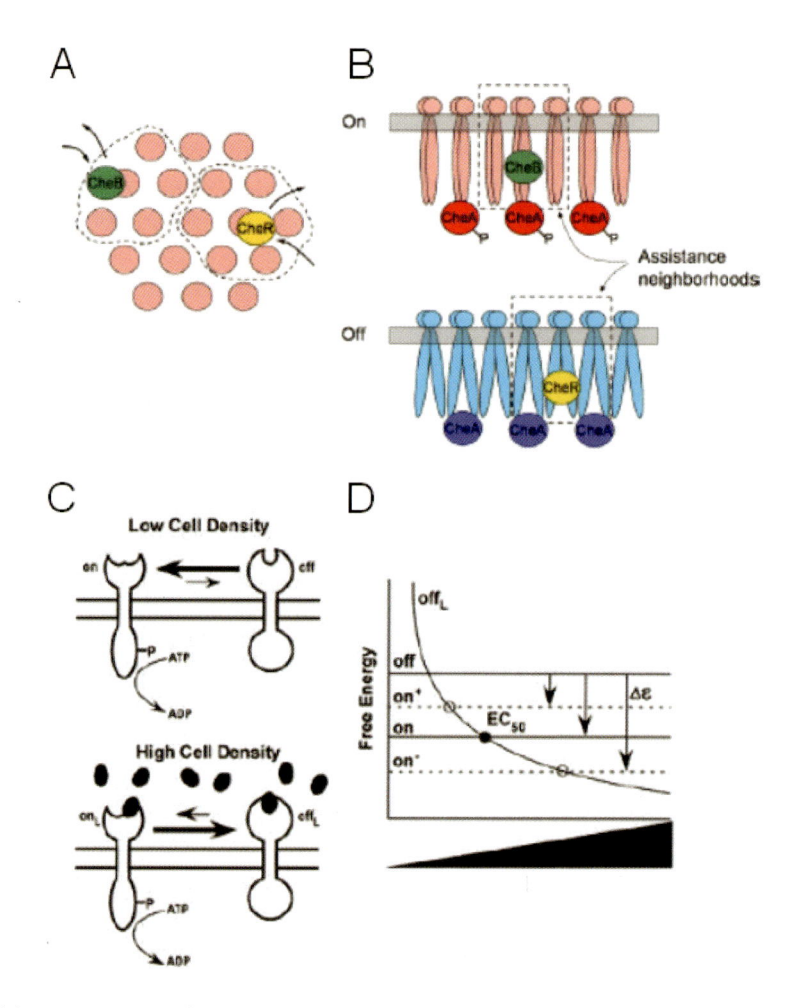

FIGURE 4-2
Chemotaxis and Quorum Sensing

The proteins responsible for eliciting both chemotaxis and quorum sensing are homologous and carry out identical biochemical reactions, yet newly acquired quantitative data reveal a stark difference in their function (see text for details). Experimental data and modeling suggest that chemotaxis receptors are poised to rapidly change signaling strength and the receptors cluster to amplify the signal. This is depicted by top (Panel A) and side (Panel B) views of receptor clusters, whereby transiently bound stimulus enzymes CheR and CheB act not only on the receptor to which they are bound but also on those receptors in their vicinity, or assistance neighborhoods. By contrast, the quorum-sensing receptors require a significant threshold signal concentration to switch activity, and they do not cluster. This is illustrated in Panel C by the single receptor and the heavy arrow, suggesting that the receptors greatly prefer the unbound state in the absence of ligand (low cell density). Panel D shows the corresponding free-energy diagram for the quorum-sensing model. SOURCES: Panels A and B, Hansen et al., 2008; Panels C and D, Swem et al., 2008, reprinted with permission from Elsevier.

sensing, information is passed internally by a series of phosphorylation events that ultimately change the activity of a DNA binding transcriptional regulator that induces/represses genes required for individual or group behaviors.

While the proteins that carry out the chemotaxis and quorum-sensing responses are homologous and, in fact, perform identical biochemical reactions, the two circuits have been shown to possess distinct design properties suggesting that the two signaling systems evolved to optimally solve very different biological problems. As a result of work by both physical scientists and life scientists, we now know that in chemotaxis, bacterial cells must respond rapidly to small, differential changes in ligand concentration. Consistent with this, chemotaxis receptors are poised to rapidly change signaling strength, spending nearly half their time in the on state. Moreover, the chemotaxis receptors cluster, which promotes signal amplification. By contrast, the quorum-sensing receptors have dramatically different signaling properties. First, in the absence of ligand, the quorum-sensing receptors are nearly always in the on state and thus require a significant threshold ligand concentration to switch off. Second, the quaternary arrangement of receptors precludes higher order complexes and thus clustering. This arrangement excludes chemotaxis-style signal amplification. Therefore, the quorum-sensing apparatus appears designed to respond slowly to the accumulation of ligand. This dramatic difference in the output of two seemingly homologous systems selects for high-sensitivity differential signaling accompanied by amplification for chemotaxis, while selecting against exactly those features in quorum-sensing signaling.

These two sensory relay systems of bacteria provide a striking example of insight into the function and design principles of biological networks that can be gained by a comparative study. They also serve as classic examples in which the close collaborations between life scientists and physical scientists, and others (biologists examining mutant phenotypes, chemists synthesizing agonist and antagonist molecules, physicists modeling network properties, and engineers studying the functioning of simplified synthetic circuits) have brought about a fundamentally new understanding of how cells process information and how cooperative behaviors evolve.

Intercellular interaction and information transfer are, of course, also central to processes of development in multicellular organisms—a subject to which we return later in discussing the problem of biological self-assembly. Indeed, the theme of specific interactions and encoding and transfer of information are ubiquitous at all levels of biological organization. It provides a natural interface with physical sciences, which can lend quantitative ideas and tools to advance our understanding of living matter. This interface is rich with fundamental questions, and we expect much progress in the near future.

DYNAMICS, MULTISTABILITY, AND STOCHASTICITY

The dynamics of simple systems are the bread and butter of physics, taught in all introductory physics courses. More sophisticated dynamical systems analysis is central to many fields in physical sciences and engineering. Dynamical systems theory provides a body of mathematical concepts and methods that allow us to describe complex dynamics in systems with many degrees of freedom (Jackson, 1991). In particular, it provides a systematic approach for identifying generic, parameter-independent behaviors as well as for reducing dimensionality by identifying the variables most essential for the dynamics. It also addresses questions of stability and multistability and provides methods for dealing with temporal changes unfolding on widely disparate timescales. Dynamical systems theory has been particularly valuable for understanding the behavior of nonlinear systems, and so the discipline has made a natural expansion into biology, the quintessential "nonlinear science."

Dynamics plays a ubiquitous role in living systems. Some of the most obvious examples are provided by rhythmic behaviors: cell cycle and circadian rhythms, respiration and heartbeat, locomotion and neural oscillations. But rhythmic behaviors are not the only manifestations of dynamics. Adaptation, growth, and evolution are also dynamical processes that cover a broad range of timescales, from seconds to millions of years. The dynamics are not always orderly and deterministic: randomness and stochasticity play important roles both on molecular and evolutionary timescales. Insight into dynamic behavior cannot be achieved without recourse to quantitative analysis. The latter provides an immediate connection to physical sciences and mathematics, fields that have developed powerful tools and concepts for studying deterministic and stochastic dynamical processes.

One common manifestation of dynamics in living systems is adaptation. Cells and organisms interact with their environment, more often than not by adaptation, a response that mitigates the effects of change. Biological examples of adaptation include the very familiar experience of our eyes needing time to adapt to darkness when the light is switched off. This phenomenon involves multiple layers of feedback operating on the molecular and cellular levels in the retina. A generally similar "negative feedback" mechanism underlies homeostasis of blood glucose level (disrupted by diabetes) and many other processes. An opposite type of feedback—positive feedback—amplifies the effect of perturbations, resulting in excitable behavior. Positive feedback plays the key role in the generation of action potentials in neurons, which was elucidated in the classic work of biophysics by Hodgkin and Huxley in 1952. Positive feedback is often associated with the existence of alternative steady states and the possibility of switching between them. Epigenetic phenomena that control cell differentiation provide another example of this behavior. In fact, it is characteristic of living systems to possess many possible states, a situation that in mathematics and physics is associated with multistability. In this case, the state of the system depends on the past history of its dynamics, meaning that the system has memory. Physical sciences provide many elegant examples of systems in which multistability and memory play important roles in dynamics, and many ideas and mathematical approaches have been developed to accurately describe such situations.

An important set of ideas bridging living and physical systems pertains to relaxational dynamics, energy landscapes, and fluctuations. Thermodynamically stable states are the minima of (free) energy: Near-equilibrium dynamics tends to drive systems into these states, while thermal fluctuations oppose this convergence. Relaxational dynamics generalizes to a broad class of nonequilibrium processes and plays an important role in the way we understand the stability and robustness of many living systems. It is also central to control theory and to the design of complex engineered systems (Freeman and Kokotovic, 2008). Evolved or designed energy landscapes can be used to understand the control of protein structure and dynamics (Figure 4-3) (Onuchic and Wolynes, 2004), while multistable dynamical landscapes in a system of interacting neurons provide a compelling hypothesis for the nature of memory (Figure 4-4) (Hopfield, 2007).

Energy landscape ideas provide powerful insight into the dynamics of complex biological molecules that catalyze chemical reactions or, in the case of molecular motors, perform mechanical work. Similar to man-made machines, "molecular machines" are built from many heterogeneous components whose interactions are carefully orchestrated. Understanding the design principles behind these machines remains a key issue for theoretical biological physicists. An excellent example of these issues is provided by a molecular motor kinesin. Our understanding of the motility of kinesin motors has advanced since the discovery of kinesin's unidirectional transport of cellular organelles along the microtubule (Vale et al., 1985). For example, force-adenosinetriphosphate (ATP) velocity relationships measured via single-molecule assays (Schnitzer et al., 2000) and kinetic ensemble experiments have enabled scientists to decipher the phenomenological energy landscape of kinesin motor dynamics. Importantly, the ideas pioneered in these studies are not restricted to the kinesins.

Many other molecular components related to cellular function, indeed, are molecular motors that utilize molecular fuels such as ATP, oxygen, cyclic adenosine monophosphate (cAMP), guanosine 5'-triphosphate (GTP), and Ca^{2+} to form a cycle of conformational switches, in the process performing work essential to maintain cellular life. It will be fascinating as new physical developments (such as nonequilibrium fluctuation theorems) shed light on the type of nanoscale dynamics that seem to have been captured so efficiently by evolution of these magnificent machines.

The ideas of multistable landscapes are also relevant in a very different context: that of formation and maintenance of memories in the brain. A range of experi-

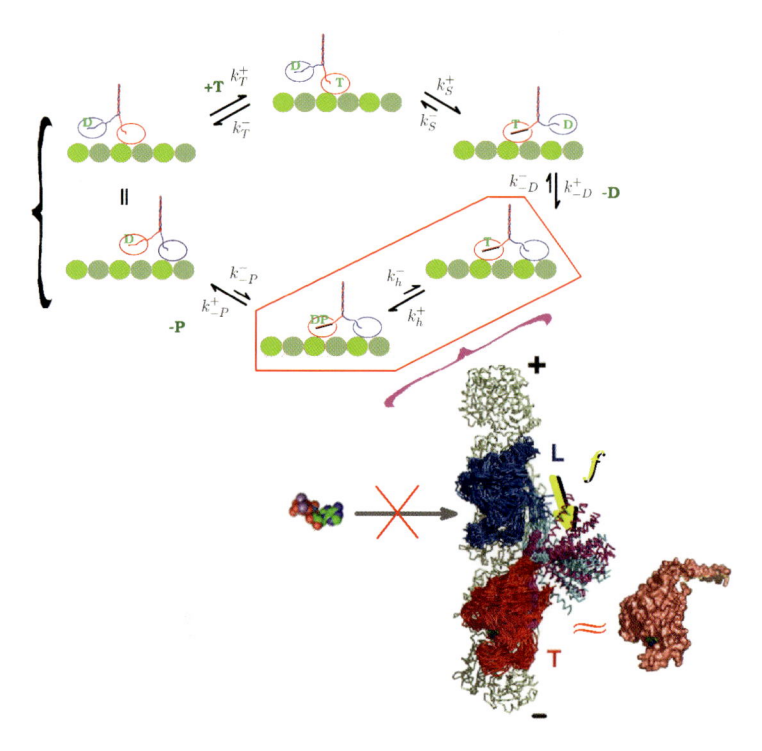

FIGURE 4-3
Dynamics of Molecular Motors

The dynamics of molecular motors directly couples protein folding to protein function. The prototypical kinesin motor, which ferries cargo throughout the cell, consists of a pair of motor domains ("heads") coupled by a flexible "neck" to a coiled-coil "stalk." The kinesin steps along the microtubule filament in a hand-over-hand fashion. Motor action occurs through an ATP-powered mechanochemical cycle (shown in the figure) involving sequential binding and unbinding of the motor heads to the microtubule along which the motor "walks." What is the mechanism coordinating large changes in the conformation of different domains—the physical "steps" that the motor takes—with the chemical activity that powers the motion? Recent modeling efforts have approached this problem using the landscape ideas for protein folding and have identified the critical role of the intermediate state (highlighted in the figure) during which both motor heads are transiently bound to the microtubule. The topological constraint of such binding introduces mechanical stress into the structure and results in an asymmetric straining of the two motor heads. Therefore the variation of kinesin structure itself is indispensable for the coordinated dynamics of the motor. The internal tension built on the neck-linker (yellow arrow) prevents the premature binding of ATP molecule to the leading head by deforming its catalytic site from its nativelike environment. The tension on the neck-linker is maintained as long as both head domains of the motor are bound to the microtubule surface. This asymmetry provides the directionality and, ultimately, the high processivity of kinesin motor action. The role of internal stress in protein conformation dynamics is likely to be central to the function of many if not all molecular machines and, as a subject of study, provides a direct example of the link between physics, chemistry, and biology. SOURCE: Changbong Hyeon and José Onuchic, University of California at San Diego.

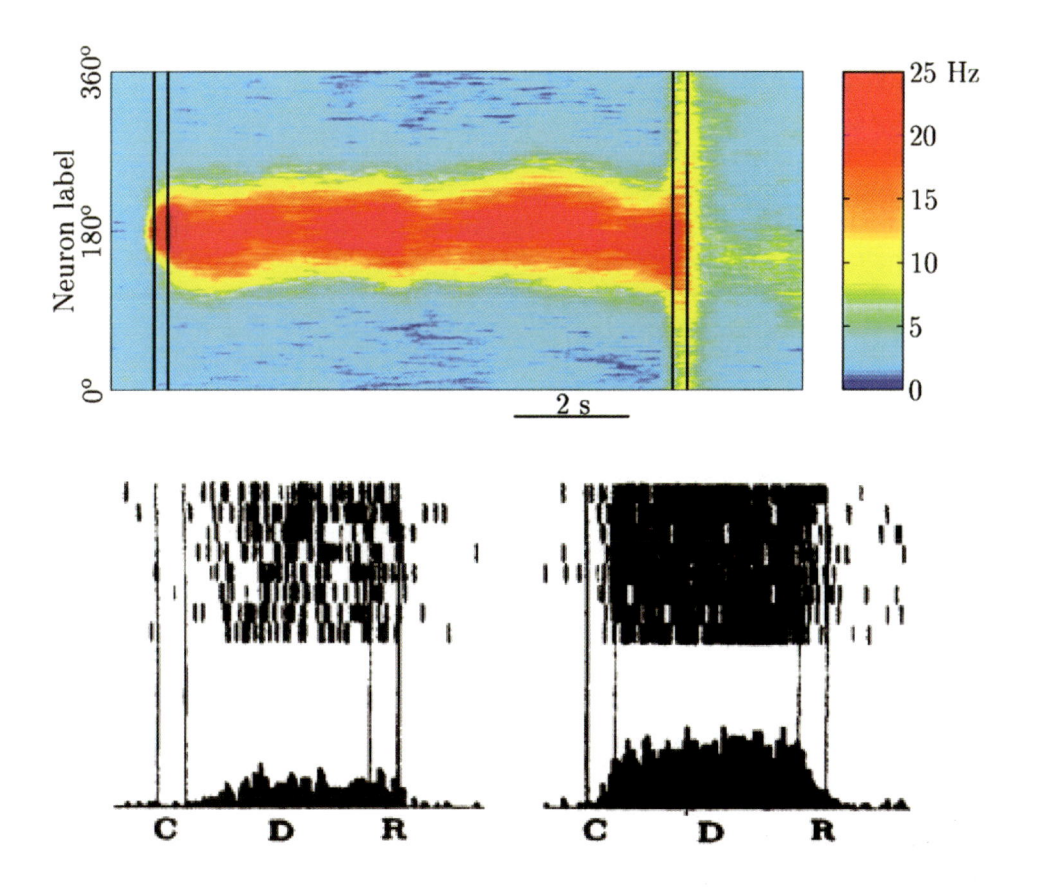

FIGURE 4-4
Brain Dynamics and Memory

Multistability in cortical networks has been proposed as a mechanism for spatial working memory. The top panel of the figure shows the spatiotemporal map of action potential firing rate in a computational model of monkey prefrontal cortex. The model network is presented with a cue at a 180-degree angle at time C and retains a memory in the form of a stable "bump" state during a delay period D until the memory is recalled at time R. The action potential firing pattern and spike rate histogram of an individual model neuron in the network (bottom right) closely resemble those of a cortical neuron in a monkey performing a cue-delay-recall task (bottom left). Action potential firing patterns in the bottom plots are represented as rastergrams, where each line shows the activity recorded during one experimental trial and each action potential is represented by a tick mark. SOURCE: Compte et al., 2000, reprinted with the permission of Oxford University Press.

mental, computational, and theoretical results suggest that memory relies at least in part on multistability phenomena. Multiple stable states can coexist in neuronal systems at several levels, from the molecular machinery at individual synapses to large-scale network activity states, and states can be stable on widely different timescales, from a few milliseconds to an entire lifetime. The basic mechanism is that a neural system becomes attracted to one of multiple stable states, which can be thought of as local minima in an energy landscape. An example is spatial working memory in a cortical network (Compte et al., 2000) where physical models reproduce the results from experiments in which monkeys are presented with visual cues at different locations (0-360 degrees) along a circle on a screen and are required to remember that location for a delay period after the cue disappears. The monkey subsequently makes an eye movement to the cued location. Recordings of electrical activity from prefrontal cortical neurons during this task show that the spike-firing rate increases in a subset of neurons when the cue is presented and remains high throughout the delay period, thus appearing to represent a memory of the cue. A computational (theoretical) network model consisting of thousands of excitatory and inhibitory neurons whose synaptic connection pattern is mimicking the cellular and synaptic organization of prefrontal cortex reliably reproduces such memory-related activity states (see Figure 4-4). This example shows that the synaptic architecture of neuronal systems can support multistability as a memory mechanism; it also illustrates the power of computational modeling and system analysis tools from the physical sciences when applied to complex biological systems.

A very important common concept equally relevant to the dynamics of physical and biological systems is stochasticity. The study of stochastic (i.e., random) behavior in biological systems provides an excellent recent example of scientific progress driven by the cross-fertilization of ideas from different disciplines, aided by technological advances enabling quantitative measurements. With the advent of fluorescent reporters and the capacity to image and individually track large numbers of single cells over extended periods of time, it became possible to observe and quantify fluctuations in different genes in single cells, even in bacteria, the smallest cells on Earth. The design of the experiments to study fluctuating biological behavior and the interpretation of the data were built on the fundamental understanding of stochastic processes in physics and mathematics. The results of the studies are shedding light not only on the molecular biology of the underlying processes but also on the possible importance of stochasticity in cell "decision making" (Suel et al., 2007; Losick and Desplan, 2008).

Studies of stochasticity show that genetically identical organisms can exhibit substantial phenotypic differences. During the process of gene expression a single gene is transcribed by RNA polymerase to produce about one hundred to several hundred mRNAs, each of which is translated to produce tens to thousands of proteins. The small number of molecules involved in this process, especially at the transcription step, can lead to substantial fluctuations in protein abundance between genetically identical cells. Because protein concentrations control the rates of many biochemical processes, stochastic fluctuations in transcription have the capacity to induce differences in physiological states. Stochastic effects of this type are, for example, suggested to play a role in tripping the switch between the dormant and active replication modes of viral infection in bacteria (Arkin et al., 1998). Elowitz and collaborators (2002) pioneered an elegant experimental design and analysis framework to measure variability, or "noise," in gene expression and to distinguish between variability that arises from intrinsic or extrinsic noise sources (see Figure 4-5). Their insight was to realize that if they used two fluorescent reporters, the difference in the levels of two reporters defined the noise intrinsic to the process of expression of the two genes while the common component of fluctuation represented the extrinsic noise arising from variability that affects the expression of all genes in a cell. Using this methodology Elowitz et al. demonstrated that substantial intrinsic noise exists in bacterial gene expression. Since this landmark study many groups have utilized this approach to measure noise in gene expression in other microbial and animal systems as well as to understand cellular individuality (Ghaemmaghami et al., 2003; Kaufmann and van Oudenaarden, 2007).

To illustrate the vast range of dynamical phenomena and the vast range of tem-

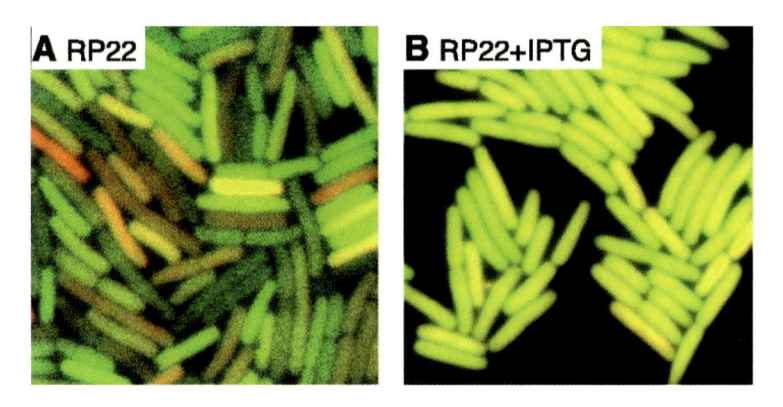

FIGURE 4-5
Molecular Noise and Stochasticity of the Cell

Single-cell imaging of genetically encoded fluorescent proteins (CFP and YFP, shown in orange and the darker green color channels) enables quantitative measurements of molecular noise in *E. coli*. Panel (A) exhibits strong fluctuations, which appear when genes (in this case CFP and YFP) are expressed at low levels because of repression by the LacI transcription factor. In panel (B), repression is eliminated and both fluorescent proteins are expressed at higher levels and the cells exhibit reduced noise. SOURCE: Adapted from Elowitz et al., 2002; reprinted with permission from the American Association for the Advancement of Science.

poral and spatial scales on which dynamics may unfold, we conclude this section with some important questions concerning the dynamics of ecosystems, biomes, and the Earth system. All of these systems exhibit properties that depend on the physiology, biochemistry, and development of individual organisms, but particular characteristics of the systems emerge depending on the interactions among organisms or interactions between organisms and their environments. These so-called emergent properties have many dimensions, and they link evolutionary, ecological, and Earth-system processes in a network of interactions that operate on diverse timescales. Emergent properties play central roles in determining the suitability of a habitat and have contributed to Earth and its life-support systems.

Nitrogen, one of the key building blocks of life, is but one example where mutually dependent interactions on many scales create an amazingly complex network, the understanding of which will require applying knowledge from both the physical and life sciences. Although nitrogen is the most abundant element in the atmosphere, the low reactivity of N_2 gas makes the atmospheric pool generally unavailable and contributes to widespread constraints on plant and animal growth by biologically available nitrogen (Vitousek, 2004). On the other hand, excess nitrogen can be a source of profound problems. Excess biologically available nitrogen plays a role in the dead zone at the mouth of the Mississippi and other large rivers. In the atmosphere, nitrous oxide (N_2O), a gas released from fertilizer, contributes to the greenhouse effect about 300 times as much, on a molecule by molecule basis, as carbon dioxide. The flow of biologically available nitrogen through the Earth system has been dramatically increased by human activity, having more than doubled in the last two centuries.

Every transformation in the global nitrogen cycle at every level of organization depends on interactions between physical and biological processes, with implications that feed back to alter the characteristics of the biological and physical environment. At the cellular scale, the assembly of nitrogen-fixing symbioses involves a complicated set of biochemical, physical, and physiological interactions between plants and micro-organisms, leading to the assembly of multiorganism "factories" that provide the delicately regulated conditions necessary for high rates of N_2 fixation. The roles of climate, nutrients released from rocks, pH in the soil, and enzymes cast into the external environment all highlight the role of the physical environment in regulating this process. Unknown variables in the area of nitrogen transformations include the importance of microbial diversity, the flexibility of the environmental constraints, and the interaction of large-scale, climate-driven processes with small-scale microbe-driven processes.

As with many of the great challenges in the environmental end of science at the interface of the biological and physical sciences, the unknowns in the global nitrogen cycle involve processes that interact across a vast range of temporal and spatial scales. Understanding nitrogen fixation or gas loss depends on effectively integrating processes that range in temporal scale from seconds for the chemical signaling among plants and micro-organisms to millennia for the evolution of microbial genomes and the development of soil texture. Relevant spatial scales range from organism to organism signaling at the scale of microns to climate change at the global scale. The committee reiterates that some of the grandest challenges in science involve understanding how emergent properties link physical and biological properties, from the molecular to the global scale, and how their properties can be managed to achieve specific effects.

SELF-ORGANIZATION AND SELF-ASSEMBLY

In physics, crystalline structure is explained in terms of energy and thermodynamics, and the dynamics of crystallization can create a multitude of forms, as, for example, in snowflakes. Assembly is another fundamental issue in the natural sciences that, like the issues of interaction and dynamics discussed above, brings complexities not encountered in the physical sciences.

Life itself is the ultimate paradigm of self-organization. What does it take to transition from inanimate to animated, living matter? The answer to this question involves issues of complexity in interacting systems, energy and information fluxes, memory, and other ingredients of self-organizing and self-replicating systems. Theoretical foundations for thinking about self-organizing and replicating systems have been laid down by von Neumann and Burks (1966). Yet, unlike Schrödinger's vision of aperiodic solids as the seat of molecular memory and genetic heritability (Schrödinger, 1944), which materialized when the structure and function of DNA were uncovered, von Neumann's ideas about self-organization have yet to bear fruit. Biological systems provide material that is ripe for applying these ideas and for the development of new ideas. Continuing interdisciplinary efforts to explore artificial life systems are likely to yield deep insights into the question of the origin of life itself.

Even the narrowest interpretation of "self-organization" as biological self-assembly and development encompasses a wide range of fascinating and important phenomena, ranging from the self-assembly of multiprotein nanoscale structures such as viral capsids and bacterial flagella to the growth and differentiation of animal tissues in the process of organogenesis. These phenomena involve dynamical processes that can be superficially compared to the execution of a program by a computer, yet all of the program "instructions" and contingencies are resident in the components themselves. How do the slight variations in viral capsomer proteins encode the structure of the capsid? How does the assembling flagellum "know" when it reaches the correct size? What are the mechanisms that ensure the proper size and proportions of a fly wing?

Lessons from biological self-assembly will advance our ability to build artificial nanostructures and advanced materials. Conversely, answers to the general questions of engineered self-assembly, such as How much information must be encoded in a set of parts to assemble a desired structure with high fidelity? may help to uncover some of the fundamental principles of living matter.

A fascinating example of molecular self-assembly in biology is provided by the bacterial flagellar motor (Chevance and Hughes, 2008) (see Figure 4-6). This remarkable molecular machine is a proton motive force (PMF) driven motor that rotates the flagellum, a flexible filament that propels swimming bacteria. Its assembly involves the coordinated sequential expression of over 50 genes encoding different protein components (Kalir et al., 2001). Protein products of genes expressed early initiate the self-assembly process and build an integral membrane ring structure, which will become the rotor, followed by the assembly onto this ring of the PMF-driven secretion system, which plays the key role in the timely export of the specific proteins that assemble the rod-and-hook structure. The assembly follows the same strategy as construction work on a skyscraper: The partially completed structure acts as a conduit for transporting building materials up to the moving "front" of construction at the top level. As the rod-and-hook structure is completed, the specificity of the secretion system switches and the structure exports protein components for the flagellar filament. We now have tantalizing insights into the molecular mechanisms that control the assembly process. For example, the length of the hook appears to be controlled by a "molecular ruler" (Journet et al., 2003) protein, which trips the specificity switch of the export system once the rod-and-hook assembly reaches a certain length. Another feedback mechanism, export of a specific transcriptional inhibitor, couples completion of the rod-and-hook assembly to the initiation of the late gene expression program (Kutsukake, 1994). The ability of scientists to dissect a molecular process of this complexity is truly inspiring and sets the stage for addressing fundamental questions concerning biological self-assembly. Can we understand evolutionary connections between different but related molecular machines? Can we distill the general principles of biological self-assembly and implement them to construct artificial nanostructures of increasing complexity? Progress in this field will clearly require a joint multidisciplinary effort among biological scientists who have a detailed understanding of these natural systems and physical scientists who are developing nanoscaled structures and advanced materials with some self-assembly or self-repair capability.

The diversity and complexity of physical forms in the living world is striking. Shapes are the most readily observable phenotypes, and it is not surprising that it was the observation of finch beak shapes that inspired Darwin's theory of natural selection. Nearly 100 years ago Darcy Thompson took the first steps to apply mathematics to the description of living shapes, but the real progress in understanding morphogenesis came with advances in genetics and molecular biology that brought us to the point where we now know the genetic factors that control the shape of

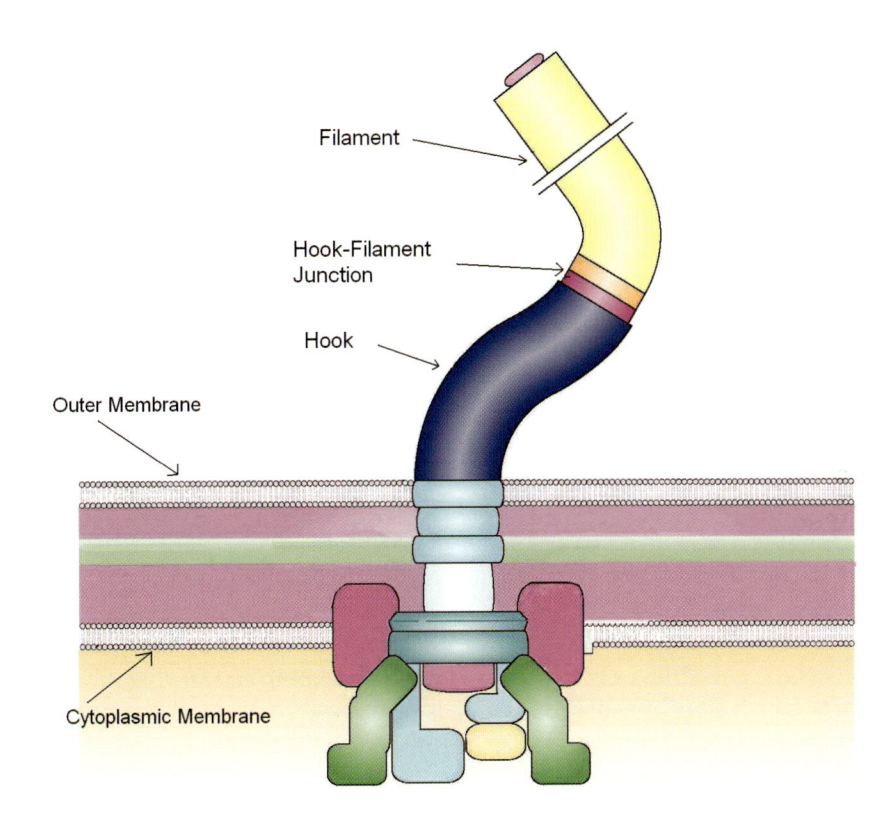

FIGURE 4-6
Flagellar Motor Self-Assembly

The flagellar apparatus is assembled within the cell wall with the filament extending outside the cell. Self-assembly of this complex structure is aided by the just-in-time expression, production, and secretion of the numerous protein components as they become required for the assembly. The assembly is regulated by multiple feedback loops that control the order of protein secretion. Switching off the supply of components determines the length of the rod, the hook, and the filament shown in the figure. Quantitative characterization of biological self-assembly in model systems such as the flagella will uncover the mechanisms of assembly of complex structures and their evolution. SOURCE: Adapted from Chevance and Hughes, 2008; reprinted and adapted with permission from Macmillan Publishers Ltd.

bird beaks (Abzhanov et al., 2004). Nontrivial structure and shape of organs, limbs, and whole bodies are the products of carefully controlled cell proliferation, rearrangement, and migration processes that constitute the developmental program of organisms. These fundamental aspects of morphogenesis are controlled by the genetically determined molecular interaction networks of the type mentioned above. Yet, morphogenesis ultimately involves the physical organization of tissues.

To relate molecular factors to macroscopic phenotypes such as the shape and structure of organs, one must understand the dynamical mechanisms and intercellular interactions that bridge the molecular and whole-organ scales. Understanding these complex phenotypes will require increasingly complex and quantitative hypotheses and the development and adoption of new quantitative tools and ideas. The problems of morphogenesis link molecular genetics with the field of pattern formation studied by physicists and applied mathematicians (Turing, 1952; Cross and Hohenberg, 1993) (Figure 4-7) and constitutes an exciting scientific frontier where interdisciplinary collaboration is likely to be the main driver of progress.

Layered onto shapes are patterns. Some familiar ones include spiral patterns of seeds in a sunflower or spines on a pineapple. Leonardo da Vinci and Johannes Kepler were among the first to ponder the remarkable mathematical regularity of these patterns: The number of visible right- and left-handed spirals is invariably a pair of successive Fibonacci numbers (5-8, 8-13, 13-21, etc.) (Adler et al., 1997). Understanding the logic of the phyllotaxis pattern has remained a challenge since that time. In 1870, Hofmesiter formulated the mathematical rules that describe phyllotaxis. More recently, Levitov showed how phyllotaxy can arise as the lowest energy state in a system of particles with long-range repulsion (1991), while Douady and Couder devised an experimental demonstration showing that sequential addition of ferrofluid droplets (which repel each other) as a function of the rate of addition naturally reproduces the full range of possible phyllotaxis patterns (1992) (Figure 4-8). Yet, the connection with plant morphogenesis was missing: What was the relation between the physical packing of repelling objects and the process of plant growth that produced beautiful patterns in biology?

The answer had to wait until molecular genetic and fluorescent imaging techniques were developed, thereby allowing investigators to determine the distribution and transport of auxin, a plant hormone promoting leaf primordia growth in the shoot apical meristem (SAM). Specifically, auxin is actively transported within the surface layer of cells by pumps that transport auxin in the direction of increasing concentration (Reinhardt et al., 2003; Jonsson et al., 2006). There is positive feedback: The steeper the auxin concentration gradient, the greater the pumping activity that further increases the gradient. This positive feedback gives rise to an instability and, in turn, to the formation of spots of high auxin concentration, which cause local outgrowth of leaf primordia. The high auxin regions forming the primordia effectively repel each other because the periphery of each high auxin region is depleted of auxin. Thus, as happens with ferrofluid droplets, a pattern, spiral in this case, emerges in the plant. To fully understand the morphogenetic implications of this molecular mechanism, plant biologists had to team up with computational modelers (Jonsson et al., 2006; Smith et al., 2006); this partnership led to a quantitative model that not only demonstrated the emergence of phyllotaxis but also made detailed, falsifiable predictions concerning the spatial

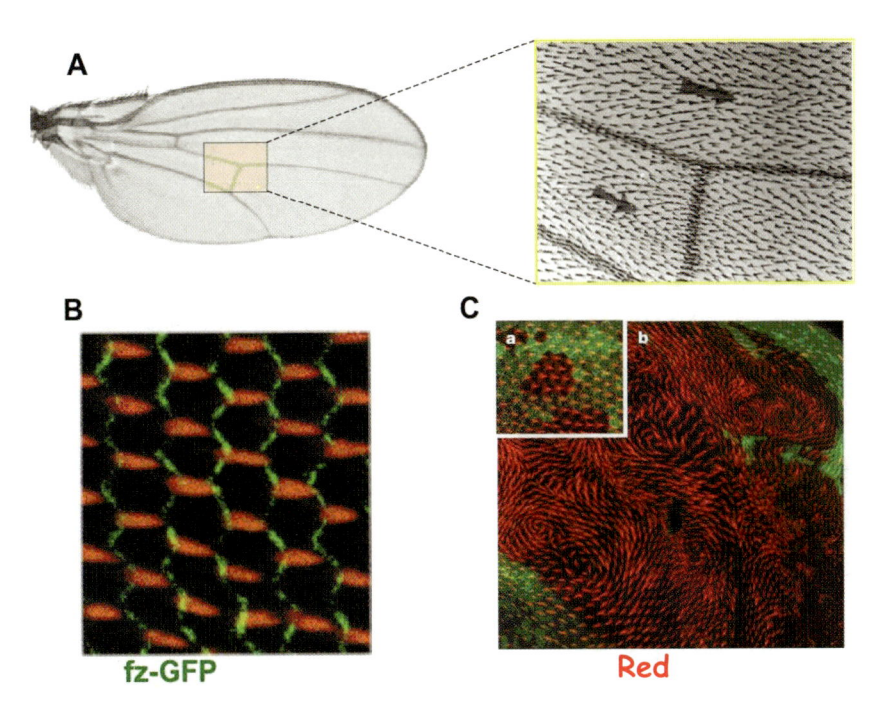

FIGURE 4-7
Planar Cell Polarity in Drosophila Wing Development

A close-up view of a fly wing (A) reveals an ordered array of distally pointing short "hairs"; (B) An array of actin bundles (orange) forming the prehairs emanates from the distal edge of each cell during the early pupal stage of wing development. Polarization of actin prehairs depends on intercellular interactions that involve the formation of asymmetric ligand-receptor complexes bridging neighboring cells. (C) Mutants defective in certain genes (in this case, a gene called Fat) show a swirling pattern of prehairs (red), which, while losing global orientation, retain local alignment. This behavior strongly suggests an analogy with ferromagnetism, which also involves formation of a globally ordered magnetization under the action of short-range interactions. We thus have a striking illustration of how physics ideas can be relevant to biological phenomena far removed from their original physical context. SOURCES: (A) and (B) Strutt, 2001, reprinted with permission from Elsevier; (C) Ma et al., 2003, reprinted with permission from Macmillan Publishers Ltd.

distribution of auxin pumps. The model for phyllotactic pattern formation is similar in spirit, although not in letter, to the Turing model in providing a bridge between the molecular and morphological scales. The phyllotactic model is now being used in close collaboration with experiment to further explore the dynamics of SAM growth and its regulation. The model also has interesting mathematical aspects that are being examined and generalized by applied mathematicians.

Ultimately, the analogy between self-assembly and developmental processes in

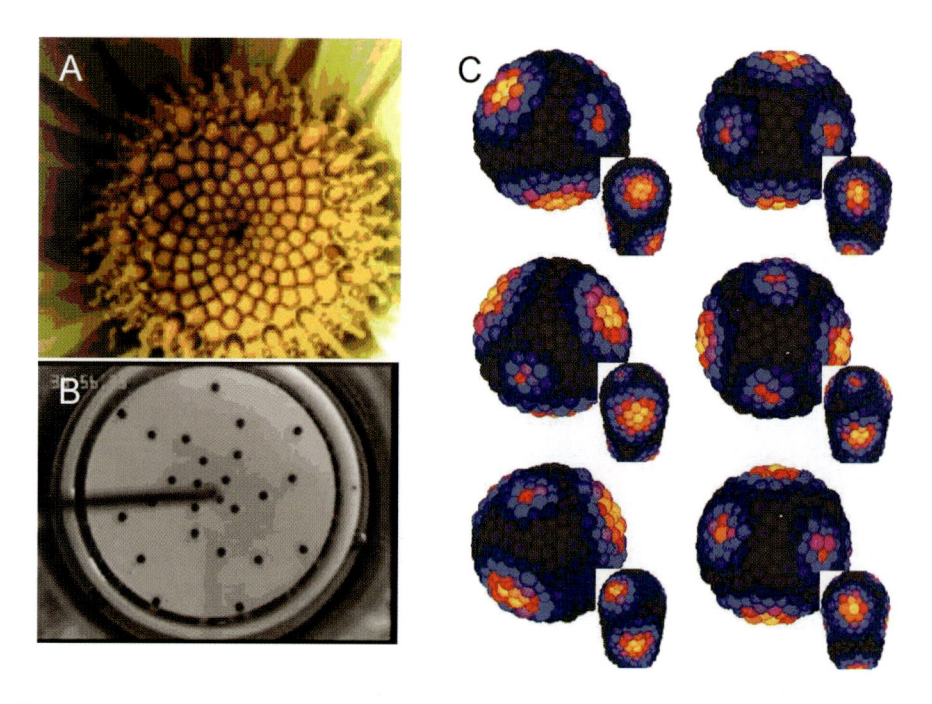

FIGURE 4-8
Phyllotaxis

Panel (A) presents a typical Fibbonacci spiral pattern of floral phyllotaxis. Panel (B) shows phyllotaxis in a physical system: a phyllotactic pattern generated by sequential addition of ferrofluid droplets in the center of a dish with a slow radial flow of viscous fluid. Panel (C) shows the early stages of the spiral formation in a computational model of auxin-driven patterning of outgrowth primordia (yellow/red) in a shoot apical meristem. SOURCES: (A) and (B) Douady and Couder, 1992, reprinted with permission from the American Physical Society; (C) Jonsson et al., 2006, courtesy of the *Proceedings of the National Academy of Sciences*.

biology suggests that not only is there no limit to the complexity of assembly, but also that only the ideas and concepts underlying the different assembly protocols remain to be discovered.

CONCLUSION

This chapter outlines just a few of the systems in which biological and physical scientists have collaborated on new concepts for viewing and solving issues in their fields. Among these, the multifaceted study of interactions and information transfer, studies of dynamical behavior, and studies of self-assembly and developmental processes provide a glimpse of how interdisciplinary approaches could bring us closer to understanding the fundamental principles of nature and the process of life itself.

REFERENCES

Abzhanov, A., M. Protas, B.R. Grant, P.R. Grant, and C.J. Tabin, 2004. Bmp4 and morphological variation of beaks in Darwin's finches, *Science* 305: 1462-1465.

Adler, I., D. Barabé, and R.V. Jean, 1997. A history of the study of phyllotaxis, *Annals of Botany* 80: 231-244.

Arkin, A., J. Ross, and H.H. McAdams, 1998. Stochastic kinetic analysis of developmental pathway bifurcation in phage lambda-infected *Escherichia coli* cells, *Genetics* 149: 1633-1648.

Chevance, F.F., and K.T. Hughes, 2008. Coordinating assembly of a bacterial macromolecular machine, *Nature Reviews Microbiology* 6: 455-465.

Compte, A., N. Brunel, P.S. Goldman-Rakic, and X.J. Wang, 2000. Synaptic mechanisms and network dynamics underlying spatial working memory in a cortical network model, *Cerebral Cortex* 10: 910-923.

Cross, M., and P. Hohenberg, 1993. Pattern formation outside of equilibrium, *Modern Physics* 65: 851.

Douady, S., and Y. Couder, 1992. Phyllotaxis as a physical self-organized growth process, *Physical Review Letters* 68: 2098-2101.

Elowitz, M.B., A.J. Levine, E.D. Siggia, and P.S. Swain, 2002. Stochastic gene expression in a single cell, *Science* 297: 1183-1186.

Freeman, R.A., and P.V. Kokotovic, 2008. *Robust Nonlinear Control Design: State-Space and Lyapunov Techniques*, Basel, Switzerland: Springer Verlag, Modern Birkhäuser Classics.

Ghaemmaghami, S., W. Huh, K. Bower, R.W. Howson, A. Belle, N. Dephoure, E.K O'Shea, and J.S. Weissman, 2003. Global analysis of protein expression in yeast, *Nature* 425: 737-741.

Hansen, C.H., R.G. Endres, and N.S. Wingreen, 2008. Chemotaxis in *Escherichia coli*: A molecular model for robust precise adaptation, *Public Library of Science Computational Biology* 4: 14-27.

Hodgkin, A., and A. Huxley, 1952. A quantitative description of membrane current and its application to conduction and excitation in nerve, *Journal of Physiology-London* 117: 500-544.

Hopfield, J.J., 2007. Hopfield Network, *Scholarpedia* 2: 1977.

Jackson, E. Atlee, 1991. *Perspectives of Nonlinear Dynamics*, Cambridge, England: Cambridge University Press.

Jonsson, H., M.G. Heisler, B.E. Shapiro, E.M. Meyerowitz, and E. Mjolsness, 2006. An auxin-driven polarized transport model for phyllotaxis, *Proceedings of the National Academy of Sciences of the United States of America* 103: 1633-1638.

Journet, L., C. Agrain, P. Broz, and G.R. Cornelis, 2003. The needle length of bacterial injectisomes is determined by a molecular ruler, *Science* 302: 1757-1760.

Kalir, S., J. McClure, K. Pabbaraju, C. Southward, M. Ronen, S. Leibler, M.G. Surette, and U. Alonl, 2001. Ordering genes in a flagella pathway by analysis of expression kinetics from living bacteria, *Science*: 292, 2080-2083.

Kaufmann, B. B., and A. van Oudenaarden, 2007. Stochastic gene expression: from single molecules to the proteome, *Current Opinion in Genetics & Development* 17: 107-112.

Klug, A., 2005. The discovery of zinc fingers and their development for practical applications in gene regulation, *Proceedings of the Japan Academy Series B-Physical and Biological Sciences* 81: 87-102.

Kutsukake, K., 1994. Excretion of the anti-sigma factor through a flagellar substructure couples flagellar gene expression with flagellar assembly in *Salmonella typhimurium*, *Molecular & General Genetics* 243: 605-612.

Levitov, L.S., 1991. Phyllotaxis of flux lattices in layered superconductors, *Physics Review Letters* 66: 224-227.

Losick, R., and C. Desplan, 2008. Stochasticity and cell fate, *Science* 320: 65-68.

Ma, D., C.H. Yang, H. McNeill, M.A. Simon, and J.D. Axelrod, 2003. Fidelity in planar cell polarity signalling, *Nature* 421: 543-547.

Onuchic, J.N., and P.G. Wolynes, 2004. Theory of protein folding, *Current Opinion in Structural Biology* 14: 70-75.

Paillard, G., C. Deremble, and R. Lavery, 2004. Looking into DNA recognition: zinc finger binding specificity, *Nucleic Acids Research* 32: 673-682.

Pavletich, N.P., and C.O. Pabo, 1991. Zinc finger-DNA recognition—Crystal-structure of a Zif268-DNA complex at 2.1-A, *Science* 252: 809-817.

Reinhardt, D., E.R. Pesce, P. Stieger, T. Mandel, K. Baltensperger, M. Bennett, J. Traas, J. Friml, and C. Kuhlemeier, 2003. Regulation of phyllotaxis by polar auxin transport, *Nature* 426: 255-260.

Schnitzer, M.J., K. Visscher, and S.M. Block, 2000. Force production by single kinesin motors, *Nature Cell Biology* 2: 718-723.

Schrödinger, E., 1944. *What is Life?* Cambridge, England: Cambridge University Press.

Smith, R.S., S. Guyomarc'h, T. Mandel, D. Reinhardt, C. Kuhlemeier, and P. Prusinkiewicz, 2006. A plausible model of phyllotaxis, *Proceedings of the National Academy of Sciences of the United States of America* 103: 1301-1306.

Strutt, D.I., 2001. Asymmetric localization of frizzled and the establishment of cell polarity in the drosophila wing, *Molecular Cell* 7: 367-375.

Suel, G.M., R.P. Kulkarni, J. Dworkin, J. Garcia-Ojalvo, and M.B. Elowitz, 2007. Tunability and noise dependence in differentiation dynamics, *Science* 315: 1716-1719.

Swem, L.R., D.L. Swem, N.S. Wingreen, and B.L. Bassler, 2008. Deducing receptor signaling parameters from in vivo analysis: LuxN/AI-1 quorum sensing in *Vibrio harvey*, *Cell* 134: 461-473.

Turing, A.M., 1952. The chemical basis of morphogenesis, *Philosophical Transactions of the Royal Society of London Series B-Biological Sciences* 237: 37-72.

Vale, R.D., T.S. Reese, and M.P. Sheetz, 1985. Identification of a novel force-generating protein, kinesin, involved in microtubule-based motility, *Cell* 42: 39-50.

Vitousek, P., 2004. *Nutrient Cycling and Limitation: Hawai'i as a Model System.* Princeton, N.J.: Princeton University Press.

von Neumann J., and A.W. Burks, 1966. *Theory of Self-Reproducing Automata.* Champaign, Ill: University of Illinois Press.

5

Enabling Technologies and Tools for Research

INTRODUCTION

One could argue that it has been the tools of the physicist and the chemist that have driven the life sciences forward at an ever-increasing rate. From the invention of X-ray crystallography to the invention of the gene chip, new technologies and tools have allowed us to look deeply into biology at increasing depth and breadth. These tools have enabled the study of the structures and dynamics that drive biological systems, and the progress has been spectacular. However, there still is much to be learned at all length scales of biological systems, from nanosized organisms to global ecosystems, and suitable tools and technologies will be critically important in studying those systems over the next 20 years.

As we ask increasingly probing questions, often guided by theory, no doubt the truly transforming ones will be those that are least expected. This chapter is not a laundry list of all the latest and greatest technologies transferring from the physical sciences to the life sciences—there is not room enough, and such a list would inevitably be skewed toward the fields of the writer. Rather, the chapter highlights areas of promising research at different size scales and seeks to identify the new technologies most urgently needed to make advances in these fields.[1]

---

[1] The recent NRC report, *A New Biology for the 21st Century*, describes a number of examples where foundational technologies have the potential of driving new scientific questions and enabling rapid technological advances; in other words, letting the problems drive the science (National Research Council, 2009).

PHYSICAL BASIS OF MOLECULAR RECOGNITION

Molecular recognition is arguably the single-most-important molecular process. It is the key to the structure-specific association of a macromolecule (protein or nucleic acid) with another molecule and is the basis for a number of subcellular activities. These include protein-ligand binding, catalysis, the action of receptors, the formation and operation of mechanical structures in the cell, the generation of energy and vectorial movement of charge, and sensing. The processes involved in molecular recognition are the same as those in the folding of proteins or nucleic acids and, somewhat more loosely, in the formation of lipid bilayers. Arguably, molecular recognition is more fundamental than any other single process in the cell—that is, more fundamental than replication and translation of DNA, synthesis of proteins, or the operation of signaling networks—since it is the basis of all of them. And, astonishingly, it still is not well understood at the molecular level.

Chemistry and biophysics have led to a picture of molecular recognition, a metaphor for which is a lock and key. In this metaphor, two molecules associate when they have complementary shapes. Complementary shapes maximize van der Waals interactions and make it possible to associate complementary electrostatic charges. Much of molecular recognition is ascribed to the hydrophobic effect, which is the association of nonpolar surfaces in water. The problem with this attractively simple metaphor is that, like many metaphors, it gives a distorted picture. Close complementary fit between associating molecules may improve the enthalpy of interaction, but it is unfavorable entropically. A better metaphor is now believed to be a cow in a tent—that is, a loose fit between molecules that minimizes the Gibbs free energy of association. What is needed for good binding of molecules to one another is the right kind of sloppy fit, but the meaning of "right kind" is not clear.

The problem of molecular association has been clearly posed for 50 years but there is still no resolution. For example, it still is not possible to rationalize quantitatively the Gibbs free energy of the binding of ligand–protein pairs or to predict the structure of new ligands for a protein, even if one has detailed knowledge of the structure of the binding site. One thing that has made the problem so difficult is that while water is clearly a necessary component of molecular recognition, the role of solvent in biology is sufficiently inconvenient that it has been ignored for the most part. For example, it is incorrect to express the molecular recognition problem in terms of protein and ligand. Instead, it must be expressed in terms of protein, ligand, water, and, perhaps, other components of biological media as well.

Understanding the interactions between water and proteins is a problem that will require high-resolution structural methods; new theoretical methods, including new methods in statistical mechanics that can handle the large numbers of particles involved; and thermodynamic analysis. Each type of information will need to be supported by physical tools such as high-resolution X-ray and neutron sources, fast computers, quantum-mechanically based potential functions, and statistical methods for single-molecule analysis.

These tools and technologies reflect a top-down approach to probing living systems: designing and building technologies using macroscale techniques and then using those technologies to examine biological systems. No doubt top-down approaches to designing new tools will continue to be extraordinarily useful, but the scope of what can be accomplished through such tools is limited. Recently, bottom-up technology—whereby self-assembly of nanostructures can be used to create new materials and to perform functions that can probe biological systems—has started to allow collecting much more useful data for understanding these and other complex biological issues (Whitesides and Grzybowski, 2002).

Until recently it was not possible to control the molecules and assemblies of molecules from which the bottom-up-designed devices are composed. However, such control is now becoming possible, and although technological hurdles still must be overcome, not only will this control allow for the design and manipulation of bottom-up technologies, but also a new array of techniques developed from such technologies should provide a substantially different perspective on a given problem. For example, the effect of a controlled-design molecular assembly on the behavior of protein-ligand binding could provide immensely useful information about the mechanisms behind molecular recognition.

STRUCTURES AND DYNAMICS WITHIN CELLS

Cellular Environment

Moving beyond the fundamentals of molecular recognition, new tools and techniques will be needed to study molecules within cells, the function of cells and assemblies of cells, and larger biological systems. The complexity of the biological milieu of a single cell, and the fundamental challenges it poses for a technology that tries to probe it, cannot be overstated. Several characteristics of the subcellular milieu are particularly relevant:

- High concentration of macromolecules,
- Extreme heterogeneity of components,
- Highly organized components, from the nanoscale and on up,
- Local protein densities that may approach the densities of closely packed spheres,
- Dynamic and directed transport of components coupled with diffusion,
- Two phases of water: one behaves like bulk solvent and the other, presumably water of hydration, has very different physical properties, and

- Rapid (subsecond) and specific variation of metabolic components on the nanoscale.

All functions within a cell occur within a medium called the cytosol. This medium fills the cell, contains ca. 300 g/L of organic material, has an ionic strength of approximately 1 mole, and is extraordinarily complex. For many years, the study of biological materials took place not in a cytosol-like medium but in water or dilute buffer solutions. Biochemical and biophysical studies have concentrated on the average (ensemble average) behavior of purified protein molecules in these dilute aqueous solutions, but the unique properties of the interiors of cells dictate that the in vivo behavior of these molecules depends on their interactions, not only with water and small molecular solutes, but also with macromolecules in a very crowded neighborhood (Minton, 2006). Although enormous strides have been made in the development of single-molecule dynamics tools, driven by optical tweezers and/or single fluorophore techniques, except for a few rare exceptions these tools are still employed in dilute solutions, not in living cells.

Dilute aqueous solutions were historically used as a model medium because the tools to do single-cell studies simply did not exist originally. When molecular biology began, proteins were very difficult to obtain in pure form, and analytical systems were crude. The introduction of the Beckman Model D spectrophotometer was a revolutionary event: it enabled, for almost the first time, quantitative physical measurements in biological systems, but its sensitivity was low. Unfortunately, the media used in those early days (very dilute solutions, low ionic strengths, buffers that did not absorb in the ultraviolet), were chosen for compatibility with the crude instrumentation of that day, not for their relevance to biology or the cell. As data began to accumulate, it was convenient to continue to use these simple media, since they provided a basis for comparing data. The result now is an immense body of data collected over the last 50 years in media known to be dissimilar in almost every respect to the media filling the cell.

It is thus an active area of research to determine the relevance (and the limitations) of this ex vivo data accumulated by molecular biologists, protein chemists, and enzymologists in understanding the processes that occur inside a cell (see Rivas et al., 2004). In the future, to ensure that ex vivo experiments are relevant to the real problems of biology, chemists, physicists, and biologists need to collaborate to define a model intracellular medium and then to develop tools to probe such a complex medium. Mapping the mass of historical data to gain such increased understanding of the cytosol will require comparing the properties of relevant biological molecules and processes in the model medium and in the cytosol and then defining an adequate cytosol model system. The model system will likely vary with the particular system being studied because the cytosol in a red blood cell is markedly different from the cytosol in a lymphocyte.

Meeting this challenge will require both bottom-up and top-down tools, as described in the preceding subsection, because the modeling of the cytosol cannot be improved without the development of techniques capable of in vivo measurements such as NMR, fluorescence resonance energy transfer (FRET), and other techniques sufficiently sensitive to study single molecules. Part of this effort would require analyzing existing information to determine what are the most pertinent characteristics of the cytosol, such as its polarity, ionic strength, dielectric constant, viscosity, polarizability, free volume, and compressibility. Another part would involve determining how to model these characteristics in a simpler fluid. The involved communities would have to agree that this simplified fluid is a valid model substitute for the cytosol. Tools both molecular (hence bottom-up) and top-down in origin would be needed to physically test and then study the model cytosol. For example, confocal-imaging coherent technologies could be used to model the dynamics of the local densities of components of the cytosol as they are transported throughout the cell.

Interactions Within Cells

Understanding the cytosol is one part of the puzzle, but understanding the internal mechanisms of cells requires functional imaging of the space- and time-resolved metabolic components and their interactions. Remarkably, demands principally arising in the physical sciences are driving the development of tools that can characterize this microecology in a spatially resolved way (Yu et al., 2006). In chemistry, probes are being developed that can be used as highly specific labels, while in physics laboratories tools are being created that can separate and detect at the single molecule level different components of a single cell. One example is the matrix-assisted laser desorption/ionization (MALDI) technique: Coupled with time-of-flight mass spectrometry it can function as a rapid and sensitive top-down analytical tool (Tsuyama et al., 2008). It has the potential to obtain molecular weights of peptides and proteins from single-cell samples and to perform in situ peptide sequencing and can map peptides in cells and tissues directly. Mass spectrometry is being used in other systems-level analyses of cellular metabolism as well, and those studies are beginning to reveal how the concentrations and fluxes of small-molecule metabolite levels in cells are controlled.

Within (as well as beyond) a single cell, biology depends on macromolecular assemblies. This dependence is apparent in the molecular mechanisms underlying gene expression, signal transduction, cell migration, cell organization, and cell division. To understand and manipulate specific biological processes, it would be advantageous to be able to design and then to generate defined molecular assemblies. While scientists are adept at devising chemical syntheses of specific compounds, their ability to design specific molecular assemblies is rudimentary, and the interface between the two technologies is still poorly understood (Bertozzi and Kiessling, 2001). Synthetic macromolecules that mimic the features and functions of naturally occurring biomolecules can illuminate fundamental principles and control biological responses. For example, synthetic oligomeric compounds have been generated that, like proteins, can adopt a specific conformation such as folding on themselves like proteins. Moreover, compounds have been devised that mimic the light-harvesting properties of the photosynthetic reaction center, and agents of this type could become new sources of energy. Macromolecules have been generated to serve as agents for drug delivery, as scaffolds for cell growth or differentiation, and as therapeutic agents. The potential of synthetic macromolecules to manipulate cellular responses by other means is great and has not been explored widely.

One strategy for creating synthetic functional assemblies is to use the molecular reaction processes employed by nature. The recognition properties of nucleic acids, for instance, are being used to generate nonnatural structures and assemblies, as is shown in Figure 5-1. This example shows how understanding biology can lead to new approaches for generating materials that function in an abiotic as well as a biotic realm. Another strategy that can be exploited to generate assemblies is to create multifunctional ligands. Agents that can direct the formation of multiprotein complexes and/or control the localization of multiple proteins within a cell would be valuable. Such ligands, which can be used like small molecules for temporal control, could illuminate how proteins assemble or how protein localization controls cellular responses. For example, multifunctional ligands could serve as scaffolds to effect signaling pathways not known to exist in nature or could endow cells with unexpected plasticity. Because modular protein assemblies are essential cellular control elements, myriad possibilities exist for using multifunctional ligands to manipulate cellular responses.

Examining Structures Within Cells

There are many underexplored but essential internal structures in the cell that exist between the atomic scale resolution of x-ray crystallography and the diffusion-limited 100-nm-scale resolution of conventional optical microscopy: cytoskeleton components, moving chromosomes, lipid rafts, folding membranes, etc. While great strides have been made in breaking the 100-nm resolution length scale of conventional microscopy, many of the techniques are quite slow (a minimum of 1 minute scanning time on a fixed sample) and require fluorescent labeling techniques, which can be difficult to implement. For example, one of the great challenges at this size range is determining how nanoscale molecular motors work. Fluorescence probes such as FRET have given us some insight into the conformational motions of these wonderful motors, but at the cost of bulky and difficult

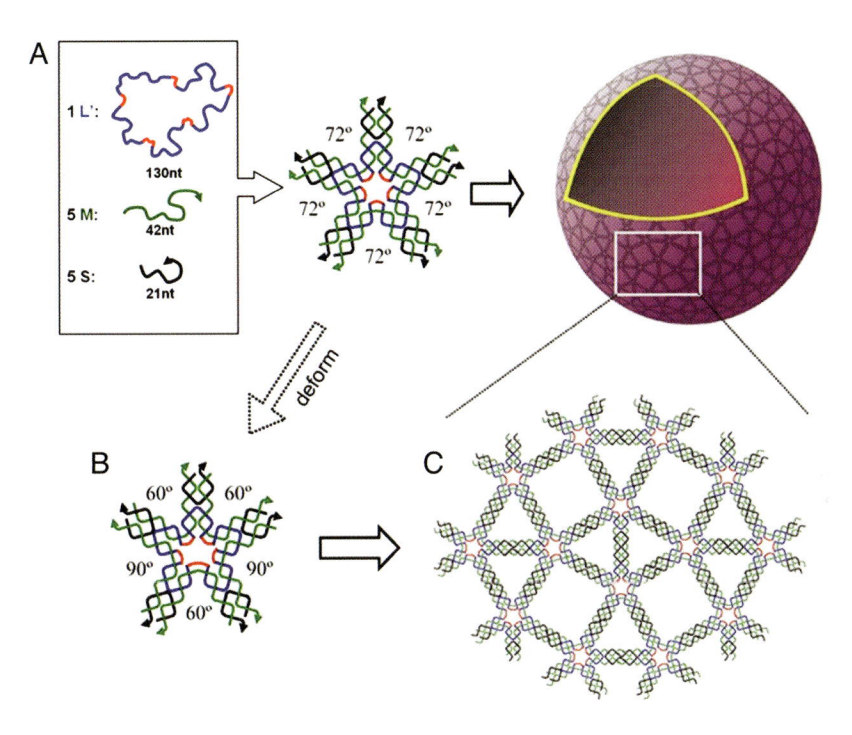

FIGURE 5-1
Synthetic Functional Assemblies

A recently proposed assembly process for creating three-dimensional nanosized objects. (A) At high DNA concentrations, five-point-star tiles can assemble, arranged into tetragonal two dimensional crystals. (B) Three-dimensional spheres are capable of self-assembling by taking advantage of the propensity of such tiles to have out-of-plane bending and asymmetrical bends in the molecular plane. (C) In the assembled structure, the angles between two neighboring branches varies (three 60-degree and two 90-degree angles), and all are different from the 72-degree angle in the free tiles. In these kinds of three-dimensional structures the conformational flexibility of the molecules is a critical part of the energetics of self-assembly. The structures created through such self-assembly processes are not only interesting in and of themselves but could also serve as encapsulation agents, nanoreactors, or organizational scaffolds. SOURCE: C. Zhang et al., 2008, courtesy of the *Proceedings of the National Academy of Sciences*.

probes that are chemically attached to the protein. Optical tweezers have allowed us to monitor the motion of these with remarkable precision, but little information is gained about how the motion actually proceeds. Molecular dynamics is greatly limited in its time range and is forced to greatly simplify the interactions between the atoms. We need tools that can see, at the single-molecule level—that is, at the sub-nanometer scale—how biological molecules proceed in functional activities, without the use of probes.

Super-resolution techniques are being developed that overcome some of the time constraints, allowing researchers to map the trajectories of individual molecules and organelles in live cells. These techniques typically involve illuminating the sample with patterned light, collecting the low-resolution image that contains moiré-fringed patterns of the sample, and then drawing from those fringed patterns high-resolution information about the sample. Different combinations of lenses, light sources, and modulated patterns have produced a plethora of acronymed microscopies, such as photo-activated localization microscopy (PALM) and stimulated emission depletion microscopy (STED), which now are allowing the mapping of trajectories of individual molecules and organelles in live cells.

Coherent, soft x-ray light sources also show great promise for studying biological systems and will probably provide high-resolution imaging in the middle of these underexplored but biologically critical length scales (Gibson et al., 2003). Furthermore, a rapid temporal sequence of images of a single cell would for the first time provide direct visualization of the complex dynamics that occur on the 10- to 200-nm length scale. Contrast mechanisms exist for high specific imaging modalities because of the narrow bandwidth of the coherent beam and its continuous tunability. Element-specific imaging (by tuning to different element L-edges and/or by using XANES/EXAFS spectroscopy in the imaging, which give the chemical state and the nearest neighbor distances and coordination, respectively) will greatly aid in the interpretation of the images. For example, we could examine how bacteria segregate their chromosomes during bacterial cell division or ask if there are specific highways for transport of mRNA molecules from the eukaryotic nucleus to the cytoplasm. Biological membranes are exceedingly important, and detailed knowledge of membrane features at the 10- to 100-nm length scale is critical. Ion channels in membranes could be imaged directly using resonances associated with specific ions. Imaging the membrane and cellular trafficking through the membrane may be the single most important application. Because over 50 percent of drug targets are G-coupled receptors or ion channels that are associated with cell signaling pathways and transport across the membrane, the pharmaceutical industry will benefit greatly if we can image receptor targets inside live, wet cells, by using specific ions at their L-edges.

Temporal or spatial relationships among individual molecules as they move within the cell cannot be captured by examining isolated static structures in vitro or by analyzing indirect biochemical or genetic data. Imaging organelle structure in frozen or red cells gives information about cellular context but is limited by its static snapshot view. Dynamic imaging of molecules in vivo is required to track structural changes over time and to obtain direct information about native structures within the cell. Despite the increasing demand to image cellular processes, however, the tools and reagents are not well developed. The linear accelerator coherent light source (LCLS) under construction and the proposed energy-recovery linear accelerator light sources (ERL) will produce coherent hard X rays, offering stroboscopic atomic-scale imaging. These light sources will revolutionize X-ray imaging and related coherent applications, including probing complex materials dynamics by X-ray photon correlation spectroscopy (XPCS). In particular, the X-ray free-electron laser (XFEL) under construction will access dynamical processes at 1 to 0.001 picoseconds including solvent and vibrational relaxations as well as energy transfer during photosynthesis. Figure 5-2 shows how the new ultrabright free electron laser sources might provide breakthroughs in this area.

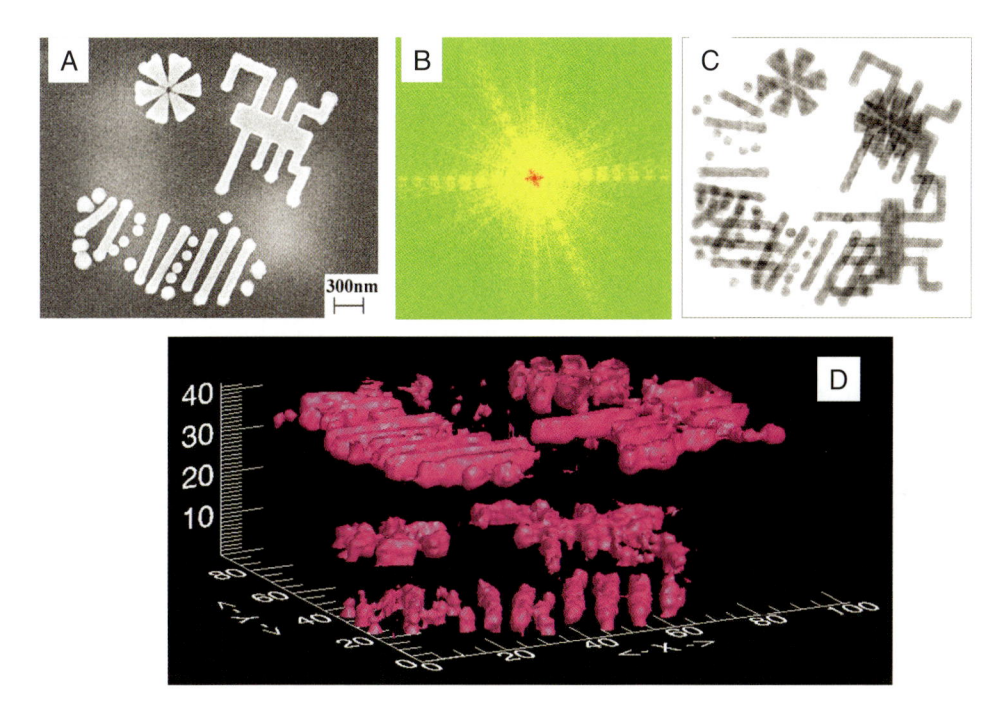

FIGURE 5-2
New Light Source Imaging

Imaging is an area where the ultrahigh brilliance and time structure of deep-UV lasers to soft X-ray, free electron lasers can have an important impact. The coherence of the source opens up imaging possibilities, including quantitative phase contrast imaging. The time structure allows for very rapid exposures over timescales below the characteristic damage time structure of some samples. This has implications for both condensed matter and biological samples. (A) A scanning electron microscope (SEM) image of a three-dimensional nanoscale specimen, which shows the surface structures but not the internal structures; (B) a coherent X-ray diffraction pattern from (A); (C) an image reconstructed from (B), showing both the surface and internal structures; (D) iso-surface rendering of a three-dimensional image reconstructed from 31 two-dimensional diffraction patterns. SOURCE: Courtesy of John Miao, University of California at Los Angeles and the California NanoSystems Initiative.

There is urgent need for a new form of system engineering at the interface between the nanoworld and the macro world. While we can now create some self-assembling nanostructures over which we have some design control, interfacing these nanoconstructs to the external macro world is difficult. The bottom-up approach might be able to give rise to massively parallel, heterogeneous, nanoscale self-assembled components, but integration of these nanoscale components into higher order structures and devices that resemble what living systems routinely accomplish is as yet out of our grasp. The integration of the two approaches, termed "hybrid top-down bottom-up" (HTBP), lies at the present cutting edge of technology development (http://www.sinam.org/). To quote a recent report from the Center for Scalable and Integrated Nano-Manufacturing (SINAM) at UCLA: "HTBP combines the best aspects of top-down and bottom-up techniques for massively parallel integration of heterogeneous nano-components into higher-order structures and devices. HTBP assembles by pick-and-place the nanoscale functional components, namely nano-LEGOs, into a defined pattern (a top-down approach); then the functional molecules attached to the nano-LEGOs can start to glue the adjacent nano-LEGOs by self-assembly, thus forming a stable structure (a bottom-up approach). Depending on designed functionalities, the nano-LEGOs can be in the form of nano-wire, quantum dots, DNA, protein, and other functional entities" (Zhang et al., 2004, pp. 126-127). Figure 5-3 shows some of the progress being made in interfacing bottom-up with top-down technologies.

THEORY AND SIMULATIONS

The diverse sizes and compositions of the heterogeneous molecules synthesized in biological environments generate a multitude of correlated phenomena on time and length scales that cannot be described with present analytical and numerical techniques. The understanding of biological processes requires the development of new theoretical approaches, modeling algorithms, and accurate effective potentials that bridge these scales. Biomolecules within the cell—in chromosomes, for example—are assembled into strong structures at short length scales but organized into soft networks at large length scales (Marko, 2008). Networks of fibers give mechanical integrity permitting interactions between molecules, possibly via compositional gradients and other long-range fields. In principle, compositional gradients result from the competition of entropy, which favors homogeneous mixing of the various components, and specific and nonspecific interactions among the molecules, which favor the formation of dense systems. The understanding of these entities requires calculations of entropy in systems with long-range interactions, transport in heterogeneous media, and interactions in inhomogeneous fluctuating environments. Moreover, since the functionality of biological organizations is dictated in part by the symmetries or lack of symmetries of their components, it is imperative to understand how symmetries are generated and broken in biologi-

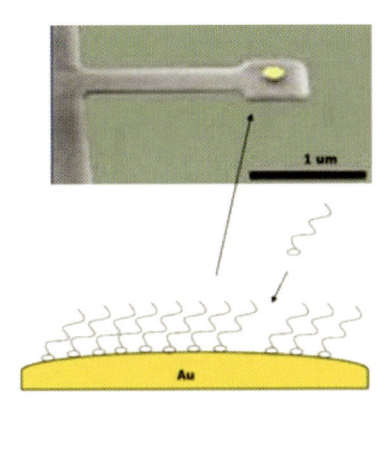

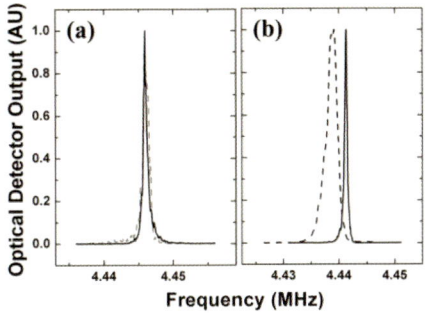

FIGURE 5-3
Interfacing Technologies

Microelectronicmechanical systems (MEMS), a top-down technology, can be combined with self-assembling monolayers, a bottom-up technology, to create extremely sensitive, label-free biosensors. A typical sensor consists of a cantilevered high-Q resonant MEMS on which the self-assembled monolayers are affixed. Here, a gold dot is placed on the cantilever to achieve adhesion for a self-assembled thiol array (top). When a molecule of the species to be detected attaches to a member of the thiol array, the resonant frequency of the MEMS shifts (bottom), thereby generating a measurable signal. Several methods for monitoring the resonant shifts and transferring the signal to external circuitry have been proposed, including direct eloectromechanical activity, wireless signals, and all-optical systems where the signal is measured using waveguide gradient forces. SOURCE: top, Harold Craigshead, Cornell University; bottom, Ilic et al., 2004, reprinted with permission from *Journal of Applied Physics* 95: 33694-33370, copyright 2004, American Institute of Physics.

cal media. Figure 5-4 shows some examples of symmetries found in biology that resemble symmetries found in assemblies of charged molecules. Finally, concepts developed in condensed matter theory to describe how emergent phenomena arise should have much to contribute to any question where interactions between many constituents is important.

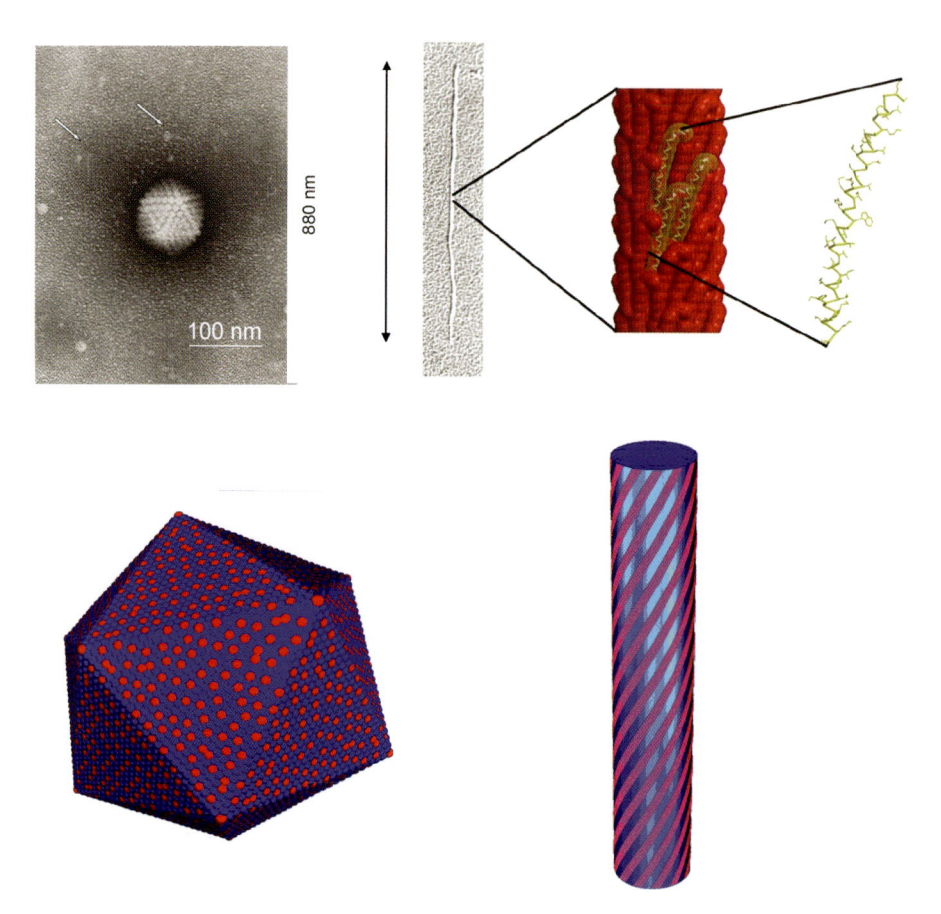

FIGURE 5-4
Symmetries in Nanostructures and Computational Challenges

The emergence and breaking of symmetries, such as rotational, translational, mirror symmetry, and chirality are known to be essential for generating functionalities at the molecular level. For instance, a conjecture has been proposed that Coulomb interactions are a means by which broken symmetries, in particular chirality, can arise at the nanoscale from the ordering of charged molecules on the surface of fibers. This may be particularly relevant in biology since in aqueous media the electrostatic energy is of the same order as the thermal energy. The figures show how basic structures and transitions induced by changes in ionic strength of co-assemblies of charged molecules, such as viruses and DNA-protein complexes, can be explored with statistical mechanics. Examples include virus structures, such as shown in the upper left, that have similar structural arrangements as obtained in models by the faceting of ionic shells into icosahedra, as seen in the lower left. Virians that assemble at bacterial membranes, as shown in the upper right, show structural similarities to the results of models computing the optimal arrangement of charged stripes over a cylindrical fiber, shown in the lower right. SOURCES: upper left, Biel, 2006; upper right, Marvin et al., 2006, reprinted with permission from Elsevier; lower left, Vernizzi and de la Cruz, 2007; lower right, Vernizzi et al., 2009, courtesy of the *Proceedings of the National Academy of Sciences*.

One approach for exploring collective behavior in physical and biochemical systems is by using complex network theory (Newman, 2008). Graphic-theoretic and statistical physics tools integrate the increasingly available information about the components of biological systems, such as genes and proteins in a cell, into a framework that can describe system-level properties. Previous network paradigms have addressed structural properties, including the determination of correlations between hierarchical structures and global properties, of complex systems. These theories can be extended to include nonlinear dynamical effects. For example, system-level analysis capable of describing and modeling the integrated functional behavior of complex systems can provide a network-based approach to control and recover metabolic function in faulty or suboptimally operating cells (Motter et al., 2008). This approach is based on identifying local modifications of the underlying network structure that can drive the system to a desired global functional behavior and can be used, for example, to identify synthetic rescues—gene pairs in which the deletion of one gene is lethal but the concurrent deletion of a second gene rescues cell viability. Besides its implications for the transformation of materials and the discovery of multitarget drugs, this approach is extremely versatile and promises to deepen our understanding of the interplay between network structure and dynamics in a variety of systems. Additional efforts that promise to lead to further advances in this field include exploratory network analysis and its variants.

The understanding of biological processes via computational methods requires force field development and coarse-grained approaches that include solvent effects within molecular dynamics, which could be developed by using physicalchemical models extended to accurately bridge length and timescales. For example, one of the main problems in computational protein folding is the question of timescales. The dynamics of individual amino acids is, in general, several orders of magnitude faster than the entire folding process of the protein. Most molecular dynamics simulations are plagued by this problem, and although preliminary efforts in this direction exist (see Sega et al., 2007), more needs to be done. Other issues in this field include the need to improve conformational search strategies for large biomolecules and a more accurate treatment of polar interactions that are crucial for identifying enzyme active sites and general properties of the protein surface. Furthermore, the role played by electrostatic interactions in shaping the folding landscape (and therefore the thermodynamics and kinetics of the folding) is far from being understood. The problem stems from the computational challenge in simulating electrostatic effects in complex environments (different dielectrics, boundary conditions, polar molecules) correctly and efficiently. The problem is still waiting for some new, and probably revolutionary, progress. Electrostatics is also crucial in the study of large structures. For studies of processes at large timescales, algorithms that accurately treat rigid-body dynamics and combine molecular dynamics with continuum boundary conditions methods may be needed. Theory is essential to construct both these and other algorithms required to describe complex biological processes and environments. Modeling will aid our understanding only if the algorithms are developed using appropriate physical arguments and mathematics.

COLLECTIVE DYNAMICS

As has been emphasized in previous chapters, the behavior of an individual cell does not determine the behavior of a collection of cells. Studying the dynamics of single cells underlies studying the collective dynamics of tissue, the extraordinarily complex assembly of cells and connective components that results in the creation of high-order plants and animals. Certainly no structure is more complex and mysterious than the human brain. Just the interconnect complexity of the human brain is truly staggering and dwarfs any foreseeable implementation of the Internet. Since each neuron has on the order of 10^4 interconnects, and there are on the order of 10^{12} neurons, the number of neuronal interconnects in the brain, which is a three-dimensional system, is on the order of 2^{51014}, a staggering number. The dynamics of these interconnections presumably results in the phenomena of consciousness, yet we have precious few tools to probe noninvasively deep into tissue with high spatial resolution. What is the enabling technology that can probe at the micron scale in centimeter depths within tissue?

In the interest of brevity, only two potentially enabling technologies are discussed here: (1) two-photon fluorescence imaging and (2) magnetic resonance imaging (MRI). Both technologies are top-down techniques and have their strengths and weaknesses. Two-photon imaging's greatest strength lies in its potential spatial resolution of submicrons within tissue. However, two-photon imaging requires the use of externally applied fluorescent probes or engineered cells that express fluorescent proteins. Because such visible light scatters, it limits penetration depths to well under a centimeter under optimal conditions. MRI's great promise is that it can do whole-tissue three-dimensional imaging, yet it is limited in its spatial resolution to, at best, the millimeter length scale. This constraint is due to Gibbs ringing, which is an inevitable artifact in MRI caused by truncating k-space. A corollary to MRI is functional MRI (fMRI), which uses changes in the metabolic state, such as oxygen concentration within tissue, to develop a spatial image of metabolic activity in whole tissue. This is a truly transformative technology, with applications from oncology to brain activity, and it is the closest thing we have to a technology that allows deep imaging of tissue. However, it is not yet good enough; the sensitivity of MRI is notoriously low due to the small magnetic moment of the nuclei; the spatial resolution is coarse; and the present cost of superconducting MRI magnets is prohibitive.

These technologies are still at the development stage, with complementary strengths and weaknesses, and in some sense all have fatal technology flaws that prevent them from achieving the spatial and temporal information we need from whole tissue analysis. It is difficult at this point in technology development to see what future technologies could be possibly developed to provide the answers we seek. The committee believes there is a powerful need to push for new enabling technologies for deep-imaging of tissue connectivity, metabolism, and dynamics at the 10-cm depth scale and at the 100-micron (at least) length scale.

Figure 5-5 shows an example of the remarkable three-dimensional imaging now possible using diffusion tensor imaging (DTI).

COMPLEX COMMUNITY SIGNALS AND SHARED RESOURCES AT LARGE LENGTH SCALES

So far, this chapter has only discussed the tools available for studying biology at the subcellular, cellular, and organism levels. Biological communities exist at many different length scales, however, from the local interactions between cells in tissue all the way up to the massive forests and plains that cover our land masses and the marine communities that exist in the oceans. These communities are in constant chemical communication with one another and reacting to the signals being sent and the production and consumption of metabolites, yet we lack the exquisitely sensitive and selective tools needed to understand the flow of chemical information among their inhabitants. These signals have profound importance: Organisms release metabolites in order to live, and they not only signal for cooperation but also present potential targets of opportunity for predators and parasites. The connection to evolution and fitness is clear, as is the connection to bioterrorism of a natural kind caused by humans.

For example, the coastal marine waters that encircle the continents are an extraordinarily important and imperiled ecology. Fully 60 percent of the world's population lives within the coastal zone (100 km from the coastline), and about 20 percent of the world's food comes from the sea. In spite of the fact that 75 percent of the Earth's surface is covered by water, most of the ocean resources come from the far smaller coastal waters.

The coastal marine ecosystem is under increasing stress that is out of proportion to the stress experienced by other critical ecosystems as the world's population continues to grow. Developing nations have historically drawn many of their resources from the sea and the coastal marine environment. This exploitation of the sea is increasing as the population grows, predominantly in large coastal urban areas. Developed nations, such as the United States, find themselves looking increasingly to the coastal marine environment for resources. Yet, in spite of this ever-increasing emphasis on a small resource and its exploitation, we are also seeing dramatic changes in the coastal marine ecosystems because of our failure to understand and protect this region. There is an urgent need to develop satellite imaging technologies that give us detailed temperature, cell density, metabolite concentrations, dissolved oxygen levels,

FIGURE 5-5
Three-Dimensional Diffusion-Tensor Imaging

Diffusion tensor imaging (DTI) is an MRI method that takes advantage of the ease with which water diffuses in various types of tissue, directly reflecting the internal fibrous structure of that tissue. By compiling many MRI measurements of the polarized water in a biological sample and extracting information from those images about the rates at which water is diffusing and its preferred directional paths, tensor mappings of the water diffusion can be developed and used to generate very detailed images of the tissue being measured. The figure visualizes a DTI measurement of the human brain. Depicted are reconstructed fiber tracts that run through the midsagittal plane. Especially prominent are the U-shaped fibers that connect the two hemispheres through the corpus callosum (the fibers come out of the image plane and consequently bend towards the top) and the fiber tracts that descend toward the spine (blue, within the image plane). SOURCE: T. Schultz, University of Chicago.

and the like of organisms and organism interactions in the marine environment. These top-down active and passive imaging technologies are increasingly powerful but still probe only a small range of the key processes and locations. Figure 5-6 shows satellite imaging of biological dead zones in the Gulf of Mexico. These zones are the result of an overabundance of nutrients from fertilizer run-off, followed by hypoxia, the condition in which bottom water oxygen concentrations are less than 2 mg/L, causing what is known as eutrophication.

Researchers have many tools and techniques at their disposal with which to study biological systems. These tools allow for the study of cells, organisms, and

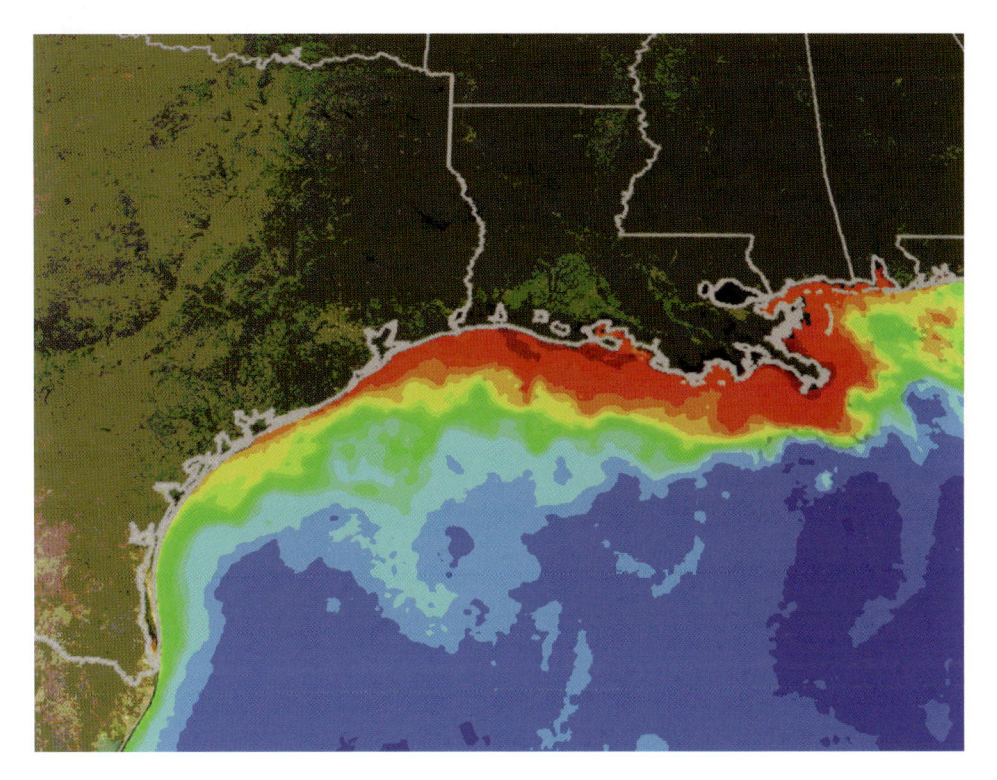

FIGURE 5-6
Satellite Imagery of Eutrophication along the U.S. Gulf Coast

Summertime satellite observations of ocean color from MODIS/Aqua show very turbid waters, which may include large blooms of phytoplankton extending from the mouth of the Mississippi River all the way to the Texas coast. When these blooms die and sink to the bottom, bacterial decomposition strips oxygen from the surrounding water, creating an environment in which it is very difficult for marine life to survive. Reds and oranges represent high concentrations of phytoplankton and river sediment. Image taken by NASA and provided courtesy of the NASA Mississippi Dead Zone Web site. Available at http://serc.carleton.edu/microbelife/topics/deadzone/general.html.

ecosystems in great detail. It is obvious, however, that new techniques must be developed to study interactions at small and large scales. The top-down approach to instrument design and technique development will continue to be important for research; indeed, for some systems and size scales, it will likely be the only path available. On the molecular level, however, bottom-up technologies promise to make the direct study and control of subcellular interactions possible.

REFERENCES

Bertozzi, C.R., and L.L. Kiessling, 2001. Chemical glycobiology, *Science* 291: 2357.
Biel, S., 2006. Adenoviridae: Human adenovirus C. *ICTVdB Management 00.001.0.01.001. Human adenovirus C. In ICTVdB—The Universal Virus Database, version 4.* C. Büchen-Osmond, ed. New York, N.Y.: Columbia University.
Gibson, E. A., A. Paul, N. Wagner, R. Tobey, D. Gaudiosi, S. Backus, I. Christov, A. Aquila, E.M. Gullikson, D.T. Attwood, M.M. Murnane, and H.C. Kapteyn, 2003. Coherent soft x-ray generation in the water window with quasi-phase matching, *Science* 302: 95.
Ilic, B., H.G. Craighead, S. Krylov, W. Senaratne, C. Ober, and P. Neuzil, 2004. Attogram detection using nano-electromechanical oscillators, *Journal of Applied Physics* 95: 33694-3703.
Marko, J.F., 2008. Micromechanical studies of mitotic chromosomes, *Chromosome Research* 16: 469.
Marvin, D.A., L.C. Welsh, M.F. Symmons, W.R.P. Scott, and S.K. Straus, 2006. Molecular structure of fd (f1, M13) filamentous bacteriophage refined with respect to x-ray fibre diffraction and solid-state NMR data supports specific models of phage assembly at the bacterial membrane, *Journal of Molecular Biology* 355, 294-309.
Minton, A.P., 2006. How can biochemical reactions within cells differ from those in test tubes? *Journal of Cell Science* 119: 2863.
Motter, A.E., N. Gulbahce, N. Almaas, and A.L. Barabasi, 2008. Predicting synthetic rescues in metabolic networks, *Molecular Systems Biology* 4: 168.
National Research Council, 2009. *A New Biology for the 21st Century*. Washington, D.C.: The National Academies Press.
Newman, M., 2008. The physics of networks, *Physics Today* 61: 33.
Rivas, G., F. Ferrone, and J. Herzfeld, 2004. Life in a crowded world, *EMBO Reports* 5: 23.
Sayre, D., 1980. *Imaging Processes and Coherence in Physics. Springer Lecture Notes in Physics*, Vol. 112, M. Schlenker et al., eds. Berlin: Springer Vertag pp. 229-235.
Sega, M., P. Faccioli, F. Pederiva, G. Garberoglio, and H. Orland, 2007. Quantitative protein dynamics from dominant folding pathways, *Physical Review Letters* 99: 118102.
Tsuyama, N., H. Mizuno, E. Tokunaga, and T. Masujima, 2008. Live single-cell molecular analysis by video-mass spectrometry, *Analytical Sciences* 24: 559.
Vernizzi, G., and M.O. de la Cruz, 2007. Faceting ionic shells into icosahedra via electrostatics, *Proceedings of the National Academy of Sciences* 104: 18382.
Vernizzi, G., K.L. Kohlstedt, and M.O. de la Cruz, 2009. The electrostatic origin of chiral patterns on nanofibers, *Soft Matter* 5: 736.
Whitesides, G. M., and B. Grzybowski, 2002. Self-assembly at all scales, *Science* 295: 2418.
Yu, J., J. Xiao, X.J. Ren, K.Q. Lao, and X.S. Xie, 2006. Probing gene expression in live cells, one protein molecule at a time, *Science* 311: 1600.
Zhang, C., M. Su, Y. He, X. Zhao, P. Fang, A.E. Ribbe, W. Jiang, and C. Mao, 2008. Conformational flexibility facilitates self-assembly of complex DNA nanostructures, *Proceedings of the National Academy of Sciences of the United States of America* 105: 10665.
Zhang, X., C. Sun, and N. Fang, 2004. Manufacturing at nanoscale: Top-down, bottom-up and system engineering, *Journal of Nanoparticle Research* 6:125.

6

Enabling Research at the Intersection: Promoting Training, Support, and Communication Across Disciplines

A great deal of substantive science is being done now where the life and physical sciences intersect, and even more transformative research is on the horizon. While the potential benefits for society are profound, realizing that full potential will require significant changes in how we educate, train, and support those undertaking this research.

The historically rigid department structure at universities, the programmatically isolated stove-piped nature of much federal funding, the different ways in which life science and physical science research are organized, and the largely separate spheres of communication that isolate life scientists from physical scientists serve as barriers to the multidisciplinary connections highlighted in this report. Intentionally or not, our system pits one scientific area against another in competition for a limited pool of resources. To obtain the benefits from research at the intersection of the physical and life sciences, it will be necessary to overcome these obstacles and to create a scientific structure that truly reflects the scientific needs and opportunities of twenty-first-century science.

Breakthroughs occur when scientists from a variety of disciplines either individually or collaboratively work on important interdisciplinary and multidisciplinary problems. Therefore, we need a new generation of scientists with both rigorous disciplinary training and the ability to communicate and work easily across disciplines. This chapter addresses the third task in the committee's charge—namely that it explore ways to enable and enhance effective interactions between the life and physical sciences, discussing some of the important areas where change is needed so that the full potential from research in the area can be realized.

CONNECTIONS BETWEEN DISCIPLINES

The degree of connection that can take place in work between disciplines varies, and the demands on those participating in and supporting that work will likewise differ. This section discusses several categories of connections; even though any such categorization is approximate at best. Actual crosscutting research efforts are more a part of a continuum rather than belonging in clearly separate categories.

The first degree of connection is simply a collaboration of experts from different disciplines, with each contributing expertise from his or her own field without crossing over into the other field ("multidisciplinary"). A physicist might build an imaging device for her neurobiologist collaborator, or a computer scientist might analyze complex sequence data generated by his geneticist collaborator. Researchers participating in this type of collaborative effort do not require extensive in-depth knowledge or skill sets from their collaborator's discipline. Although always desirable, truly interdisciplinary training in this type of minimally cross-disciplinary work is not strictly necessary. However, the collaborators must at least speak each other's language—that is, communicate across disciplinary borders.

A second type of research is conducted by individuals who were originally trained in one of the classical disciplines but since have acquired skills and knowledge in another discipline. This might include, for example, additional coursework at the graduate level or postdoctoral training in the other field. The research performed by these individuals, either by themselves or with others similarly trained, is currently often referred to as interdisciplinary research. Such cross-training mostly happens owing to the interests and initiative of individuals. Work of this type could be facilitated by encouraging classically trained physical and life scientists to transcend their disciplines and acquire education and training in other fields. One mechanism that fosters the cross-over of individuals trained in the physical sciences into the life sciences is discussed in the next subsection.

A third category is research in new fields that have emerged from previous connections between disciplines ("interdisciplinary integration"). Such fields as biomedical engineering and biostatistics combine features of several traditional disciplines into a new discipline. The emergence of such fields is not new, as several of the distinct fields now firmly established in today's universities, including molecular biology and biochemistry, have their origin in the intersection of past disciplines.

Culture of Separation between the Life and Physical Sciences

Conducting research at the intersection of the physical and life sciences requires bringing together not only separate disciplines but also, in many senses, separate cultures. While biologists, physicists, and chemists may not be that different in some respects, the way they conduct, communicate, and organize their science can be very different in other respects. Building the interdisciplinary and multidisciplinary connections discussed throughout this report will require overcoming these distinctions so that scientists from a variety of disciplines can work together on problems of common interest.

The heart of biological research has been single-investigator-initiated projects of relatively short duration. Although some of the recent projects, such as the Human Genome Project, have been large ones, they are not the main source of support for biology. In contrast, much federal support for large segments of the physical sciences, such as astrophysics and high-energy physics, goes to large facilities and programs, and those projects, unlike a typical life sciences project, almost certainly have permanent staff and involve large amounts of instrumentation and construction support.

Physical scientists who participate in these large-scale research efforts also are accustomed to awarding credit to large numbers of investigators and are comfortable with scientific papers that have hundreds of authors. In contrast, publications in the life sciences tend to include no more than one, two, or three principal investigators as authors, along with a handful of graduate students, postdoctoral researchers, and others who conduct the actual research. As the number of authors per manuscript increases, it becomes progressively more difficult for most of the authors to receive adequate credit for their contributions. Physics seems to have found mechanisms to circumvent this problem, but the issue of how investigators are evaluated remains one of the major cultural divides between the two disciplines.

Life and physical scientists have typically been members of largely separate scientific communities, attending different meetings and reading different journals. The committee encourages universities, professional societies, and funding agencies to seek ways to connect researchers across disciplines. The Keck Futures Initiative of the National Academies provides one model for bringing together researchers from across disciplines (see Box 6-1). Recommendation 3 in the 2008 National Research Council report *Inspired by Biology* (National Research Council, 2008b) may also be helpful in this regard, as it suggests summer courses that bring together life scientists and physical scientists and allow researchers to be introduced to other disciplines.

Another successful model of interdisciplinary community building has been developed by the Kavli Institute for Theoretical Physics and the Aspen Center for Physics. These institutions bring physicists and biologists together for extended workshops in a format that allows new collaborations to germinate. This format has been adapted from long-standing practices of the theoretical physics community but has proven very effective in the interdisciplinary setting: Most of the life scientists introduced to the highly interactive experience of these workshops choose

BOX 6-1
National Academies' Keck Initiative

National Academies has launched a 15-year effort to realize the untapped potential for research that crosses disciplinary boundaries, thanks to a $40 million grant from the W.M. Keck Foundation. The Keck Futures Initiative brings together approximately 100 top scientists from a variety of disciplines to consider a series of questions and challenges. Following several days of conversation and engagement in groups, participants have the opportunity to apply for seed grants that will enable launching ideas generated at the conference.

Several of the Keck Futures conferences have focused on research at the intersection of the physical and life sciences, including the 2003 meeting, "Signals, Decisions and Meaning in Biology, Chemistry, Physics and Engineering"; the 2004 conference "Designing Nanostructures at the Interface Between Biomedical and Physical Systems"; and the 2008 focus, "Complex Systems." Additional information about the Keck Futures Initiative is available at http://www.keckfutures.org/

to participate over and over, resulting in steady growth of the interdisciplinary communities associated with these institutions. Workshops like these influence the research agendas of the participating scientists and play an important role in developing a common language uniting different scientific communities. They lay the foundation for breakthroughs not possible when the disciplines work in isolation.

Culture and Organization of Academia

It is not only the culture of disciplines that complicates research at the intersection of the physical and life sciences but also the culture and traditions of academia. In particular, most universities are divided into colleges and then into academic departments, usually along traditional disciplinary lines such as biology, chemistry, and physics, which then form the basic administrative units for the university. The hiring, promotion, and granting of tenure for faculty, graduate programs, and "credit" given for instruction generally are all determined with respect to these departments. Although many faculty members have joint appointments in more than one department, they often must meet tenure criteria and take on teaching responsibilities in a "home" department.

Because their work often does not squarely fall within the purview of a single department, faculty members working at the intersection of the physical and life sciences may find it difficult to get support in any one department. Although a full discussion of university tenure policies is well beyond the scope of this report, the disciplinary-based nature of promotion and tenure is an impediment to multidisciplinary education and research. These statements are not intended to declare that the traditional disciplinary structure is misguided or shortsighted, but rather that alternative support structures may be called for in light of these particular challenges.

The organization of university research has, however, shown promise of reform. In particular, multidisciplinary centers organized around shared research topics or common research goals are becoming more common. These centers can provide an alternative model for organizing research activities in a way that complements—but does not replace—existing departments. Centers enable universities to move rapidly into emerging fields, provide a home for faculty working at the intersection of disciplines, and develop courses and train students free of traditional departmental and disciplinary constraints. They provide opportunities increasingly viewed as important to members of the next generation of scientists, many of whom are attracted to work on certain problems rather than in particular subdisciplines. Because these centers can achieve the multidisciplinary goals discussed in this report without replacing the existing university structure, they are an attractive mechanism for promoting research at the intersection of the physical and life sciences in the medium term. Recommendation 1 calls for further support for these efforts.

RECOMMENDATION 1. Federal and private funding agencies should expand support for interdisciplinary and multidisciplinary research and education centers. In particular, extramural funding should be provided to establish and maintain center infrastructure and research expenses. Initial (e.g., 5-year) salary support for investigators performing research that spans disciplines should also be included, with continuing salary support for faculty associated with the center provided by the host institution(s) or department(s). To support these centers, universities will need to implement multidepartment hiring practices and tenure policies that support faculty working collaboratively within and across multiple disciplines, establish shared resources, and provide incentives for departments to promote multi-departmental research and cross-disciplinary teaching opportunities.

ORGANIZATION OF SUPPORT FOR RESEARCH

Research at the intersection of the physical and life sciences necessarily falls between the boundaries of disciplines, which also means that it often falls between the boundaries of how research support is organized. Several federal agencies—including the Department of Defense (DOD), the Department of Energy (DOE), the National Institutes of Health (NIH), and the National Science Foundation (NSF)—support research at this intersection through a variety of mechanisms, from individual, investigator-initiated research grants to large centers and consortia. Which agency is most appropriate for a given proposal can be difficult to determine, because much of the work at the intersection of the physical and life sciences overlaps the interests of a variety of programs and agencies but does not fit squarely within any single funding program. The challenges exist not only between agencies, but also between divisions within an agency and between funding programs within the same division.

Such silo effects are certainly not limited to research at the intersection of the physical and life sciences but are common in any type of crossdisciplinary investigation. Hence if solutions can be identified that improve support for research at the interface between biology and physical sciences, these solutions could serve as models to improve the climate at other interfaces such as that between basic biology and medicine.

The differences in level of support among federal funding agencies, and the tendency to fund canonical research rather than research at the disciplinary interface, has led to the perception that one area is more critical than another. While federal support is indeed limited, research funding is not always a zero-sum game. The committee hopes for renewed focus on supporting and conducting the best science, including that which crosses traditional boundaries. One manifestation of this hope would be enhanced collaboration between agencies with a greater number of joint programs than at present. Another recent report makes a similar point for realizing opportunities *within* the life sciences: cross-agency collaboration is essential for supporting the needs of science and society (National Research Council, 2009).

Federal funding agencies and private sponsors of research have supported research at the intersection of the physical and life sciences for several decades. Such support usually has been for highly specific programs and almost always for programs contained within a single agency.

The NSF has played an important role in pioneering such research. Its molecular biophysics program, administered by the Division of Molecular and Cellular Biosciences, has funded research at this interface since the late 1960s and 1970s and has had a seminal impact. This program fostered the development of techniques such as X-ray crystallography and nuclear magnetic resonance before they became common, and later expanded to support research on theory and simulations. The Physics of Living Systems program, administered by the NSF Division of Physics, evolved from the earlier biological physics programs and supports scientists using the tools of physics to study biological problems at the molecular level. Of the 11 Physics Frontiers Centers, 2 have biological physics as their focus.

NIH's National Cancer Institute recently initiated a program under its Center for Strategic Scientific Initiatives. This program, which seeks to promote the types of collaborations proposed in this report, addresses outstanding issues in research on cancer. Begun in 2008, it has conducted three workshops and is proposing to fund four, five, or six centers for 5 years, at an annual budget outlay of $15 million to $20 million. The program has the potential to provide valuable guidance on the encouragement and support of cross-disciplinary work in the life and physical sciences.

Perhaps the most interesting federal support for such research has come from the Defense Advanced Research Projects Agency (DARPA). DARPA has funded innovative projects in such areas as the detection of infectious disease agents and methods for rapid development and deployment of novel therapeutics against infectious disease, a number of which involved work at this intersection. DARPA support has some features that differentiate it from other federal support. For example, it provided extramural funding directed to targeted objectives, often involving large grants over a relatively short time. Reviews were carried out at a single location with rigorous security measures. As a result, investigators seemed more willing to expose novel ideas. Reviewer workloads were lighter than for comparable study sections at other agencies, meaning that there was more time for the review panel to discuss the pros and cons of particular projects in detail. Because DARPA program managers also had considerable discretionary authority in making funding choices, agency funding priorities could be shaped by the professional staff.

High-risk research at the intersection of the physical sciences and life sciences seems to have received a more encouraging welcome from the private sector than from the federal sector. Perhaps this is not surprising as small companies supported by venture capital are accustomed to supporting potentially transformative research whose failure rates approach 90 percent. Examples of successful work at this intersection include the development of technology platforms for diagnostics and for therapeutics. Ongoing efforts include the development of new tools for the analysis or manipulation of biological systems such as high-resolution optical microscopy, and cryo-electron microscopy and tomography. Such platforms inevitably require interdisciplinary teams of biological and physical scientists, engineers, and software developers, as well as a willingness to try unproven technologies. It is impossible to conceive of most federal support mechanisms taking such risks, but the committee hopes that the federal government will devote at least a small portion of its funds to potentially transformative research at the intersection of the physical and life sciences.

In addition to the need for more funding for research at the intersection of the physical and life sciences, there is a need to assess what changes should be made in the mechanisms for administering programs that support such research so that those funds will be utilized more efficiently. How can interdisciplinary and multidisciplinary proposals be appropriately ranked in competition with single-discipline proposals? What new models are needed to support research that spans funding programs and agency boundaries?

Current support mechanisms for interdisciplinary work require that research proposals be submitted to specific funding agencies and often require acknowledging when simultaneous submissions are being made to other funding agencies. In times of tight funding it is hard for one agency's reviewers to avoid downgrading an application submitted to more than one agency in favor of a proposal that has no other possible sources of support. A simple solution would be to allow submissions to multiple agencies without prejudice during the review process—for instance, information about pending support might not be revealed in the prospectus being considered by review panels.

A more meaningful solution would be to establish crossdisciplinary funding that spans agency divisions or even entire agencies. These joint programs would need to undergo a single, cross-agency review process rather than independent, parallel-review procedures at each participating agency. One example of such a joint program is in the area of mathematical biology and is being supported by both NSF and NIH (see Box 6-2).

It is, admittedly, more difficult to evaluate interdisciplinary research proposals. The community of established reviewers whose skills are well-founded in a pair of disciplines is small. People in one discipline will usually favor that field since it is what excites them, it is what they most easily understand, and it is where they can most easily recognize rigor and innovation. And having separate reviews by people with expertise in each of the respective areas and then merging the scores will disadvantage the investigator, as each of the disciplinary experts is likely to undervalue the proposal. In the private sector, where multidisciplinary research is more common, evaluations are prolonged, individuals with many perspectives are involved, the number of proposals is relatively small, and a project is revisited and reevaluated continually. These characteristics work well but are not a solution for federally funded work, because the sheer volume of proposals would make them impractical.

The committee recognizes that successful programs for funding innovative interdisciplinary science exist within funding agencies, several of which were described above. However, opportunities are lost due to lack of cooperation between agencies to make programs that are more than the sum of the parts. Therefore, the committee calls on the White House and its Office of Science and Technology Policy (OSTP) and Office of Management and Budget (OMB) to develop standing mechanisms that will facilitate, rather than impede, interagency collaborations. In particular, in Recommendation 2 the committee calls upon OSTP and OMB to work through the National Science and Technology Council (NSTC) to establish

BOX 6-2
NSF/NIH Joint Program in Mathematical Biology

The Directorate for Mathematical and Physical Sciences at the NSF and the National Institute of General Medical Sciences at the NIH have established a joint program in mathematical biology.[1] The goal of the program is to engage practicing mathematicians in the core of biomedical research.

Proposals are reviewed by a single review panel that incorporates expertise from both the life sciences and mathematics, ensuring that candidate proposals incorporate rigorous mathematics and engage substantive biological questions. Most successful proposals demonstrate a clear commitment to substantive collaboration between one or more biologists and one or more mathematicians. The joint review panel considers review criteria for both NIH and NSF so that it is not necessary to conduct separate reviews at each agency.

Both agencies must sign off on each award, but the grants are ultimately awarded by either NSF or NIH and subject to the award requirements at that agency. The decision on which agency makes the award is at the option of the agencies, so investigators apply to a single program.

The success of the program is helped by the commitment of both NIH and NSF and the involvement of program and review staff from both agencies. The agencies have also helped to build a community of scholars in the program; they organized a meeting of principal investigators in 2003, and another is being considered in 2009.

[1] Available at http://www.nsf.gov/pubs/2006/nsf06607/nsf06607.htm

a standing interagency working group on multidisciplinary research under the NSTC's Committee on Science.

> **RECOMMENDATION 2. The Office of Science and Technology Policy (OSTP) and the Office of Management and Budget (OMB) should develop mechanisms to ensure effective collaboration and cooperation among federal agencies that support research at the nexus of the physical and life sciences. In particular, OSTP and OMB should work with federal science agencies to establish standing mechanisms that facilitate the funding of interagency programs and coordinate the application and review procedures for such joint programs. Moreover, the National Science and Technology Council should establish a standing interagency working group on multidisciplinary research within its Committee on Science, with focus on the intersection of the physical and life sciences.**

BOX 6-3
Interagency Working Group on Plant Genomes

The Interagency Working Group on Plant Genomes was established in May 1997 under the direction of the NSTC's Committee on Science to pursue crop genomics in the public sector. The group's National Plant Genome Initiative (NPGI) focused on developing genomics tools that would transition discoveries in model plants such as Arabidopsis to crop species.

The group was charged to "(1) identify science-based priorities for a plant genome initiative; and (2) determine the best strategy for a coordinated Federal approach to supporting such an initiative, based on respective agency missions and capabilities" (National Research Council, 2008a, p. 16). Since its founding, the initial group of participating agencies has expanded and the coordinated NPGI now incorporates most of the federal investments in plant genomics. Although small in overall investment, the U.S. continues to lead the world in the productivity of plant science research.

There are several successful models of NSTC working groups and subcommittees charting a coordinated research agenda, including those focused on plant genomes and global change research. As described in Box 6-3, the former provides a coordination function in the area of plant genomics. The latter subcommittee serves the coordinating body for the U.S. Global Change Research Program, which is made up of 13 federal department and agencies and includes an integration and coordination office to implement the program's strategic plan.

Following the model of the NPGI (see Box 6-3), the interdisciplinary working group on multidisciplinary research should begin by identifying all of the agencies with an interest in research at the physical and life sciences and overcoming the barriers to multidisciplinary connections pointed out in this report. The group would then be charged with developing multiagency solicitations and review procedures that would support research that falls between existing government funding programs.

SUPPORTING TRANSFORMATIVE RESEARCH

Although transformative research has tended to occur at the boundaries of existing disciplines, cross-disciplinary proposals often have difficulty surviving the review process intact. Most U.S. research funding is based on the prospective analysis of a research proposal and significant preliminary results are almost required as evidence that the proposed plan can succeed. In fact, there is often a sense that investigators must cite their past research in proposals that support their future research. In an era of limited resources, it is perhaps natural for review committees to favor proposals that are likely to succeed—that is, relatively conservative proposals that extend the boundaries of past research incrementally—and to avoid taking chances on proposals that might be transformative but that also have a high risk of failing.

The challenge of supporting high-risk, high-reward research has drawn significant attention in recent years. For example, the recent ARISE report from the American Academy of Arts and Sciences recommended that all federal agencies should have programs that focus on supporting innovation with relatively simple and rapid application processes and called for an evaluation of such programs to ensure that they are, indeed, supporting at least some transformative research (AAAS, 2008, p. 36). ARISE also recommended that funding mechanisms and review processes should "nurture, rather than inhibit, potentially transformative research" by tweaking review criteria, providing more flexibility and resources for agency program managers to support exploratory projects, and by establishing interdisciplinary review panels to consider high-risk research proposals across fields.

Not only the research community but also the funding agencies themselves have called for change to support transformative research. For example, the National Science Board, which establishes the policies of the NSF, called for an NSF-wide Transformative Research Initiative that would be distinct from other programs and allow the NSF to establish new structures and procedures for transformative research (National Science Board, 2008). Similarly, an NIH initiative on enhancing peer review has recommended that at least 1 percent of investigator-initiated research awards be directed to transformative research programs. The self-study also recommended an analysis of interdisciplinary research applications to determine how they were assigned to review and how successful they were in obtaining funding, and recommended an editorial board model for review of interdisciplinary research (National Institutes of Health, 2008).

Several ongoing programs, most of them operating outside the normal peer review system, have the goal of supporting transformative research. Each of these has clear relevance to research at the intersection of the physical and life sciences. For example, since 2004, NIH has supported "individual scientists of exceptional creativity who propose pioneering—and possibly transforming approaches—to major challenges in biomedical and behavioral research"[1] with the NIH director's Pioneer Award. Based on a special review process, the program places greater emphasis on the investigator than most and even includes in-person interviews with the finalists. While the program is too new to have been formally evaluated, it

---

[1] NIH Director's Pioneer Award overview, found at http://nihroadmap.nih.gov/pioneer/.

is striking that a large fraction of Pioneer awardees work at the interface of biology and physics and that several are physical scientists, given that NIH supports few physical scientists overall. Several NIH institutes, including the National Institute of General Medical Science, have made one round of awards for Exceptional, Unconventional Research Enabling Knowledge Acceleration (EUREKA) grants. The proposed research is expected to have a substantial impact on a significant fraction of the scientific community. As with the Pioneer award, the EUREKA application and review process emphasizes significance and innovation in addition to experimental approach and other considerations.

The Howard Hughes Medical Institute (HHMI) has provided leadership in recognizing the importance of and funding work at the physical/life sciences intersection. Since 1990, HHMI has held national competitions for investigators, who then become HHMI employees while retaining their faculty positions and their laboratory location at their university or research institute. By supporting people, not projects, HHMI rewards retrospective evidence of innovative contributions more than prospective analysis of research plans both in its initial appointment of investigators and in the 5-year reviews of their appointments for renewal. Interdisciplinary work is overtly sought and valued, and a number of HHMI investigators are chemists, physicists, computer scientists, and engineers who are tackling biological or biomedical problems. Some of this multidisciplinary and interdisciplinary work now is supported in a dedicated facility HHMI recently established known as the Janelia Farm Research Campus (see Box 6-4).

To enhance the opportunity for potentially transformative research, the committee makes recommendations designed to make proposals from more than one principal investigator (PI) more common. The National Science and Technology Council and its Research Business Models Subcommittee have advocated enhancing opportunities and providing uniformity in evaluating proposals with more than one PI. The committee feels that there is room for further enhancement and hopes that multi-PI proposals and awards will become easier to achieve because they directly promote projects with equal participation of researchers from more than one discipline.

> **RECOMMENDATION 3. Federal and private funding agencies should enhance the ability of more than one researcher to serve as principal investigator (PI) on research projects. Each PI should receive full credit for participation on the grant, with the lead PI serving as the administrative contact.**

By this recommendation, the committee does not intend to require that most or even many grant programs only provide multi-PI awards, as there is a need for the continued support of single-investigator multidisciplinary research. Rather, the optimal models for funding would include a mix of single-PI, two-PI, and multiple-PI research activities—with the particular organization dependent on the

BOX 6-4
HHMI'S Janelia Farm Research Campus

Starting in 2000, HHMI tried to discern which sorts of basic biomedical research, if any, were challenging to support through its university-based investigators program—that is, which types of research needed a new model in order to be fully realized. This analysis led to the conclusion that multidisciplinary research at universities was often frustrated by departmental decisions and prompted HHMI to build a free-standing, wholly supported multidisciplinary research institute.

The HHMI Janelia Farm Research Campus opened in 2006 in the northern Virginia suburbs of Washington, D.C. By 2010, it will house 44 research groups, each capped at a small size (six researchers in the laboratory of a "group leader," two in the laboratory of a fellow). The small group size encourages frequent interaction and collaboration and is inspired by the small teams that contributed to the success of Bell Labs.

One of Janelia Farm's initial research emphases is mapping and understanding the neural circuits responsible for all complex behavior. This area of research provides an interactive environment, bringing together a variety of investigators, including neuroscientists, physicists, computer scientists, and chemists. Additional information about Janelia Farm is available at http://www.hhmi.org/janelia/.

specific research. The critical need, at this time, is to break down the administrative barriers that prevent scientists from assembling in the most effective way to secure extramural funding and in conducting research.

The committee also recommends that federal and private funding agencies explicitly support potentially transformative research. By their nature, such programs should incorporate application and review procedures that are consistent with multidisciplinary research and incorporate the viewpoints from a variety of scientific disciplines. The committee also feels that these programs and any others for cross-disciplinary research should be continually assessed to be sure that they are meeting those goals.

> **RECOMMENDATION 4. Federal and private finding agencies should devote a portion of their resources to support potentially transformative research, including opportunities at the intersection of the physical and life sciences. These sponsors should have peer review procedures that incorporate the viewpoints of scientists from a variety of disciplines. Moreover, they should continually assess the effectiveness of these grant programs and the review procedures to ensure that they are meeting the desired aims.**

EDUCATING SCIENTISTS AT THE INTERSECTION OF THE PHYSICAL AND LIFE SCIENCES

Pursuing research at the intersection of traditional disciplines will require a workforce prepared to work at the boundaries between disciplines. This will mean changing the way we educate the next generation of scientists.

A number of reports in the last decade call for enhanced quantitative training of life scientists and say it will be a critical need for the life sciences in the future (e.g., National Research Council, 2003, 2005a, 2005b, 2008). Because research in the biological sciences is becoming increasingly quantitative, a greater ability to model biological phenomena using mathematical language is needed. Moreover, the vast collection of data now available to researchers through such fields as genomics has introduced a complexity and need for data analysis not previously relevant in the life sciences. Many of these changes have happened within the last decade, leaving even some life sciences faculty members ill equipped.

The committee considered the best mechanisms for empowering life scientists with the appropriate degree of mathematical sophistication and for providing educational settings in which students want to learn the math required to address problems of interest. The committee could have simply recommended that all students in the life sciences take one or more classes in advanced mathematics. But this would just produce students with a background in biology and mathematics but no assurance that they would be able to apply the mathematics they learn in one class to the biology they learn in another. Rather, biology and mathematics should be treated together and it would be even more effective if the physical sciences were also included. The committee is hopeful that the NSF program for interdisciplinary training for undergraduates in the biological and mathematical sciences[2] will reveal best practices that will apply to a broad range of students.

Collaborative teaching efforts by life scientists and applied mathematicians, physicists, or engineers may be the best way to provide biology students with the quantitative and problem-solving skills they need and also help to bridge the physical and life sciences in the classroom setting. Such courses can also provide students with a model of how scientists and mathematicians approach problems, demonstrating the relevance of multidisciplinary approaches and the need for mathematical sophistication.

Enabling Interdisciplinary Research Starting at the Undergraduate Level

Enhanced quantitative training for biologists is an important first step in fostering researchers who can work at this intersection, but it is only one step. It will

---

[2] Solicitation 08-510; http://www.nsf.gov/pubs/2008/nsf08510/nsf08510.htm.

also be critical to increase the exposure of physical scientists to the life sciences and vice versa.

The most basic way to realize this goal would be to have all physical science undergraduates study the principles of biology such as genetics and evolution and to have all biology majors receive appropriate preparation in physics, chemistry, and mathematics. In fact, biology majors at many undergraduate institutions have long been required to take specific courses in these other departments, although often there is no attempt to make these courses relevant to life science majors. However, this approach reinforces the impression that the scientific disciplines are discrete, isolated entities and leaves it to the student to draw connections about the relevance of other disciplines to biology.

An alternative approach, which is harder to establish but probably would have a more lasting impact, is to integrate applications, examples, and problems from other disciplines into core courses to increase relevance. For example, quantitative aspects could be given more emphasis in existing biology courses and materials from the life sciences could be incorporated into existing physical science and mathematics classes. It is just such an integration that a recent NRC report proposed (National Research Council, 2003).

More pertinent would be a single introductory course that incorporates elements of both the physical and life sciences and would introduce biology and physics majors to the basics of both disciplines. Although a small number of institutions offer a common introductory course (see Box 6-5 for an example), such experiments are far from universal. The committee encourages more institutions to design such courses while recognizing that developing such courses requires addressing a complex and at times conflicting set of goals. Any such course will need to ensure that the knowledge bases of the respective disciplines are presented coherently and at an appropriate depth. At the same time, the illustrations and thematic questions that attempt to integrate the various approaches to similar questions must be clearly presented and relate back to the original disciplines. Finally, time limitations undoubtedly will require making difficult decisions on which topics covered in traditional courses will be diminished or treated in a different way.

It also is important that these classes not be taught in isolation from the rest of the curriculum but be integrated into the set of courses offered by all involved departments. To meet these needs, faculty members from those departments must meet regularly to coordinate curriculum planning. Funding agencies also can play a role in these education efforts by not only facilitating the development of these courses but trying to evaluate the effectiveness of such strategies in increasing student familiarity and knowledge of both the physical and life sciences.

BOX 6-5
Introductory Interdisciplinary Science at the Evergreen State College

The Evergreen State College in Olympia, Washington, offers a 1-year-long interdisciplinary program entitled "Introduction to Natural Science: Life, the Universe, and Everything." The course brings together unifying perspectives from physics and chemistry to provide a conceptual and experimental introduction to natural science. It takes a thematic approach, focusing on cycles and transformations of matter and energy in both living and nonliving systems, which allows students to see similar ideas emerging at a variety of levels.

The course is team-taught and always involves a chemist and a biologist, with a third faculty member from areas such as physics, computer science, or geology. Different areas of science are integrated throughout the course, including into exams that test knowledge in more than one subject area—especially areas that bring the disciplines together.

Enrolling approximately 100 students per year, the course combines lectures, problem-solving activities, laboratories, field trips, seminars, the reading of primary research literature, and independent scientific investigations by small groups in collaboration with one of the faculty members. It serves as preparation for more advanced courses in the physical and biological sciences, as well as in the health and environmental sciences.

RECOMMENDATION 5. At the undergraduate level:
- **Universities should establish science curriculum committees that include both life scientists and physical scientists to coordinate curricula between science departments and to plan introductory courses that prepare both those who would major in the life sciences and those who would enter the physical sciences.**
- **Professional scientific societies should partner with peer societies across the life and physical sciences to organize workshops and provide resources that will facilitate multidisciplinary education for undergraduates.**
- **Federal and private funding agencies should offer seed grants to academic institutions to develop new introductory courses that incorporate both the physical and life sciences and to professional societies for organizing workshops and developing resources for multidisciplinary education. They should also support research to identify best practices in such education.**

The committee acknowledges that adding material to a curriculum or to an individual course may require institutions and faculty to make difficult decisions about which existing topics to eliminate or treat in a different way. It is beyond the scope of this report to define the specific courses and subjects for a curriculum but the NRC report *Bio2010: Transforming Undergraduate Education for Future Research Biologists* may assist institutions in considering these issues (National Research Council, 2003).

Integrating Life and Physical Sciences for Graduate Students and Postdoctoral Researchers

Undergraduate training that brings together the life and physical sciences would be an important foundation for expanded multidisciplinary connections at the graduate and postdoctoral levels. Research habits of mind and the socialization of budding scientists as full members of the research community develop during Ph.D. and postdoctoral training. Ensuring their greater involvement in research across disciplinary boundaries and regular interaction with scientists from a variety of disciplines will make them comfortable with multidisciplinary research early in their careers.

As one example of supporting graduate training in emerging interdisciplinary fields, HHMI has partnered with the NIH to establish the Interfaces Initiative, which provides training grants to institutions to develop interdisciplinary graduate programs (see Box 6-6). Despite the promising experiment begun by HHMI and NIH, most graduate programs do not offer students an easy opportunity to work with researchers across the life–physical sciences interface. Most departments offer either their own doctoral program or an umbrella program that does not span the divide between the life and physical sciences.

Federal and private funding agencies provide support for a large number of doctoral students and postdoctoral researchers in the life and physical sciences. Many of the trainees are supported as research assistants on research grants awarded to principal investigators, while others are part of institutional training grants or supported by individual fellowships. The committee sees a role for leveraging this extramural support to encourage interdisciplinary training that spans the life and physical sciences.

RECOMMENDATION 6. Federal and private funding agencies should offer expanded training grants that explicitly include graduate students and postdoctoral researchers from fields across the life and physical sciences and that require the involvement of academic departments from both the physical and life sciences. Funding agencies should also offer administrative supplements to existing research grants that would enable a principal investigator in the life sciences to support a postdoctoral researcher with a background in the physical sciences, or vice versa.

BOX 6-6
HHMI-NIBIB Interfaces Initiative

The HHMI and the NIH's National Institute of Biomedical Imaging and Bioengineering (NIBIB) have jointly developed and supported an interdisciplinary graduate research training program: the HHMI-NIBIB Interfaces Initiative. First awarded in 2005, the initiative's 4-year training grants were established with the goal of teaching graduate students to work effectively across disciplinary lines. This initiative takes advantage of HHMI's ability to catalyze the creation of new university programs and the ability of NIBIB to sustain such programs once formed.

The initiative supports institutional training grants rather than fellowships to individual predoctoral students because institutional grants require a greater degree of interaction by faculty from diverse fields. Faculty cooperation, in turn, can help drive institutional change and ongoing connections between disciplines.

Among the programs that have been supported are those with a focus in mathematical, computational, and systems biology (University of California, Irvine); multiscale analysis of biological structure and function (University of California, San Diego); and biophysical dynamics and self-organization (University of Chicago). Additional information about the HHMI-NIBIB Interfaces Initiatives is available at http://www.hhmi.org/grants/institutions/nibib.html.

The committee sees value in encouraging those who have received a doctorate in a traditional discipline, not just scientists training in interdisciplinary graduate programs, to consider applying their knowledge to research questions at the intersection of the physical and life sciences, and to give them a jump start on their research careers. One prominent program in this area is the Career Awards at the Scientific Interface program of the Burroughs Wellcome Fund (BWF), which has been offered since 2002 (see Box 6-7). The NIH has created the Pathway to Independence Award, which offers a similar two-phase postdoctoral/faculty award but with no specific consideration for physical scientists who have little background in the biomedical sciences. Career transition programs such as the BWF Career Awards and the NIH Pathway to Independence can encourage institutions to establish positions for creative early-career scientists that might not have otherwise existed. They also help young scientists to become established in their careers by reducing the funding pressure on them and fostering their transition to independent research careers (National Research Council, 2005c).

Several of the other career development awards offered by the NIH support scientists who propose to train in a new field. For example, the Mentored Research Scientist

BOX 6-7
Burroughs Wellcome Fund Career Awards at the Scientific Interface

The Career Awards at the Scientific Interface (CASI) program, sponsored by the Burroughs Wellcome Fund (BWF), offer 5 years of funding that bridges advanced postdoctoral training with the first 3 years of faculty service. The program accepts applications from individuals early in their postdoctoral careers who have a Ph.D. in mathematics, physics, chemistry, computer science, statistics, or engineering and whose work addresses biological questions. The program provides 1 to 2 years of salary and research support during the postdoctoral portion of the award, with the balance of the 5 years of support once the recipient has assumed a faculty position.

The program is unique in providing advice regarding career development to the fellows: Each awardee is counseled on the terms of the faculty offers he or she receives and is given advice on how that offer compares to others given to former fellows. The majority of the CASI fellows have gone on to top junior faculty positions and are developing into leaders in interdisciplinary fields such as systems biology and computational biology. More information about this program can be found at http://www.bwfund.org/pages/129/Career-Awards-at-the-Scientific-Interface/.

Development Award (K01) from the National Human Genome Research Institute is open to individuals with degrees in computer sciences, mathematics, chemistry, engineering, physics, and closely related scientific disciplines. The Mentored Quantitative Research Career Development Award (K25) from the NIBIB supports scientists and engineers with little or no experience in medicine or the life sciences to develop the relevant research skills that will allow them to conduct basic or clinical biomedical imaging or bioengineering research.

A number of additional programs support early-career scientists by providing an infusion of funds early in their research careers—but without an explicit career transition element. Proposals for these awards are often relatively short and frame a set of research questions, rather than a detailed experimental plan that contains significant amounts of preliminary data. The NIH director's New Innovator Award complements the NIH director's Pioneer Award (discussed in a preceeding section) and gives 5 years of funding that emphasizes innovation, with a focus on scientists who received their most recent doctoral degree 10 or fewer years ago but who have not yet received an R01 award.[3] Appropriately, this program has included a number of scientists working at the intersection of the life and physical sciences. NSF's

---

[3] More information about this program can be found at http://nihroadmap.nih.gov/newinnovator/.

Faculty Early Career Development (CAREER) program provides 5 years of support to tenure-track but untenured faculty. Among the private-sector programs that currently or historically have supported new investigators are the Markey Scholar Awards from the Lucille P. Markey Charitable Trust, the Keck Distinguished Young Scholars in Biomedical Research from the W.M. Keck Foundation, the David and Lucille Packard Foundation Fellowships for Science and Engineering, the Beckman Young Investigator Program from the Arnold and Mabel Beckman Foundation, the Pew Scholars Program in Biomedical Sciences from the Pew Charitable Trusts, the Searle Scholars Program from the Kinship Foundation, the Damon Runyon Cancer Research Foundation Scholar Award, the Sloan Research Fellowships, the McKnight Scholar Awards from the McKnight Endowment Fund for Neuroscience, and the Klingenstein Fellowship Awards from the Esther A. and Joseph Klingenstein Fund. In addition, several universities offer time-limited independent research fellowships for early-career investigators, including the Carnegie Institution for Science, the Whitehead Institute for Biomedical Research, the San Francisco and Berkeley campuses of the University of California, and Harvard University.[4]

All of these awards that occur early in a scientist's career can be critical to encouraging new faculty to take on innovative projects in multidisciplinary areas and should be encouraged. Despite these examples, however, the number of grant opportunities that support scientists switching fields of study is quite limited in both federally supported and private-sector programs.

> **RECOMMENDATION 7. Federal and private funding agencies should offer opportunities for both early-career and established investigators trained in one discipline to receive training in another and apply their experience and training to interdisciplinary problems. In particular, postdoctoral career awards should be established that facilitate the transition of a candidate prepared in a physical science field to apply that training to important questions in the life sciences and vice versa. Funding agencies should also provide expanded support for experienced investigators to receive training in a new field, perhaps in the form of sabbatical fellowships.**

This recommendation is similar to those that have been offered before in reports including those of the NRC (National Research Council, 2005c), and the National Academies (2007).

---

[4] For more information about these and other programs focused on new investigators, please see Chapter 2 of National Research Council, 2005c.

REFERENCES

American Academy of Arts and Sciences, 2008. *ARISE: Advancing Research in Science and Engineering: Investing in Early-Career Sciences and High-Risk, High-Reward Research.* Cambridge, Mass: American Academy of Arts and Sciences.

National Academies, 2007. *Rising Above the Gathering Storm: Energizing and Employing America for a Brighter Economic Future.* Washington, D.C.: The National Academies Press.

National Institutes of Health, 2008. *2007-2008 Peer Review Self-Study, Final Draft.* http://enhancing-peer-review.nih.gov/meetings/NIHPeerReviewReportFINALDRAFT.pdf.

National Research Council, 2003. *Bio2010: Transforming Undergraduate Education for Future Research Biologists.* Washington, D.C.: The National Academies Press.

National Research Council, 2005a. *Advancing the Nation's Health Needs: NIH Research Training Programs.* Washington, D.C.: The National Academies Press.

National Research Council, 2005b. *Mathematics and 21st Century Biology.* Washington, D.C.: The National Academies Press.

National Research Council, 2005c. *Bridges to Independence: Fostering the Independence of New Investigators in Biomedical Research.* Washington, D.C.: The National Academies Press.

National Research Council, 2008a. *Achievements of the National Plant Genome Initiative and New Horizons in Plant Biology.* Washington, D.C.: The National Academies Press.

National Research Council, 2008b. *Inspired by Biology: From Molecules to Materials to Machines.* Washington, D.C.: The National Academies Press.

National Science Board, 2008. *Enhancing Support of Transformative Research at the National Science Foundation.* NSB-07-32. Arlington, Va.: National Science Foundation.

Appendixes

Appendix A

Statement of Task

The committee will:

1. Develop a conceptual framework for the scientific forefronts at the interface between the physical and life sciences and conduct an assessment of this work.
2. Identify and prioritize the most promising research opportunities at this interface, articulate the potential benefits to society, and recommend strategies for realizing them.
3. Explore ways to enable and enhance effective interdisciplinary collaboration, such as education, training, instrumentation, and cyberinfrastructure, which bring together the life and physical sciences to address the most compelling opportunities.

The committee will explore areas such as biomolecular machines; gene regulation and signal transduction; mechanics and spatial structure of the cell; and the origin of self-replicating systems. The committee will also identify other key research areas in which the interaction of physical and life scientists is needed for scientific advancement.

Appendix B

Meeting Agendas

**FIRST MEETING
WASHINGTON, D.C.
SEPTEMBER 14 AND 15, 2007**

Friday, September 14, 2007

Open Session

1:30 p.m.	Perspectives from the Office of Science & Technology Policy	J.H. Marburger, III
2:00	Perspectives from the Department of Energy (15 min each)	
	Basic Energy Sciences	A. Kini
	Biological and Environmental Research	D. Drell
2:30	Perspectives from the National Institutes of Health	
	National Institute of Biomedical Imaging and Bioengineering	W. Heetderks

3:00	Break	
3:15	Perspectives from the National Science Foundation: Panel Discussion (15 min each)	
	Physical Sciences	T. Chan, Directorate for Mathematics and Physical Sciences
	Life Sciences	J. Collins, Directorate for Biological Sciences
3:45	Selected comments (5 min each)	Chemistry (Z. Rosenzweig) Materials research (D. Brant) Physics (K. Blagoev) Engineering (S. Rastegar) Molecular and cellular biology (M. Henkart)
4:10	Panel discussion	All
4:45	Perspectives from Research Corporation for Science Advancement	J. Gentile
5:30	Discussion	
6:00	Reception	
7:15	Evening lecture	K. Dill, Bridging the Sciences Coalition
8:30	Adjourn for the day	

Saturday, September 15, 2007

Closed Session

SECOND MEETING
WASHINGTON, D.C.
DECEMBER 18-20, 2007

Tuesday, December 18, 2007

Closed Session

Wednesday, December 19, 2007

Open Session

7:30 a.m.	Continental breakfast	
8:00	Welcome, charge to committee, purpose of symposium	Erin O'Shea, *Co-chair* Peter Wolynes, *Co-chair*
8:15	The energy problem and what we can do about it	Steven Chu, Lawrence Berkeley National Laboratory
8:30	Discussion	All
9:05	The role of biology in confronting the climate-energy challenge	Daniel Schrag, Harvard University
9:20	Discussion	All
9:55	Break	
10:10	Challenges in biodefense research	James Baker, University of Michigan
10:25	Discussion	All
11:00	Challenges at the intersection of physics and population biology	Alan Perelson, Los Alamos National Laboratory
11:15	Discussion	All
11:50	Lunch and concurrent cross-cutting breakout sessions (tools, training and education, culture)	
1:00 p.m.	Challenges in research at the molecular to organism level	Jay Keasling, University of California, Berkeley, and Lawrence Berkeley National Laboratory

Appendix B

1:15	Discussion	All
1:50	Biomaterials and biomimetics: Challenges and opportunities	Joanna Aizenberg, Harvard University
2:05	Discussion	All
2:40	Break	
2:55	Challenges in cognition and learning	Larry Abbott, Columbia University
3:10	Discussion	All
3:45	Challenges in research on the origin of life	Jack Szostak, Massachusetts General Hospital, Howard Hughes Medical Institute, Harvard
4:00	Discussion	All
4:35	Frontiers in fluorescence imaging	Jennifer Lippincott-Schwartz, National Institutes of Health
4:50	Discussion	All
5:25	Public comment session	
5:45	Reception	
6:30	Adjourn	

Thursday, December 20, 2007

Open Session

9:15 a.m.	Conversation with new project sponsor	Nancy Sung, Senior Program Officer, Burroughs Wellcome Fund (by teleconference)

Closed Session

9:30 a.m.

5:00 p.m.   Adjourn

THIRD MEETING
WASHINGTON, D.C.
APRIL 29-30, 2008

Tuesday, April 29, 2008

Open Session

12:00 p.m.	Training the workforce: The Burroughs Wellcome Fund experience	Nancy S. Sung, Senior Program Officer, Burroughs Wellcome Fund

Closed Session

1:00

Wednesday, April 30, 2008

Closed Session

FOURTH MEETING
BERKELEY, CALIFORNIA
JULY 29-31, 2008

Tuesday, July 29, 2008

Closed Session

Wednesday, July 30, 2008

Closed Session

Thursday, July 31, 2008

Closed Session

Appendix C

Biographies of Committee Members

Erin K. O'Shea (NAS), *Co-Chair*, is professor of molecular and cellular biology and director of the FAS Center for Systems Biology at Harvard University, where she is a member of the biophysics faculty. She is also an investigator for HHMI. Dr. O'Shea received her Ph.D. in chemistry from MIT in 1992. Before teaching at Harvard, Dr. O'Shea was a member of the medical school at the University of California, San Francisco. In 2004, she was elected to the National Academy of Sciences for her critical contributions to our knowledge of how cells sense and respond to their environment. She has been chair of the Committee on Degrees in Chemical and Physical Biology at Harvard.

Peter G. Wolynes (NAS), *Co-Chair*, is the Francis Crick Chair of the departments of chemistry and biochemistry and of physics at the University of California, San Diego. He received his Ph.D. in chemical physics from Harvard in 1976. After a postdoc at the Massachusetts Institute of Technology, he returned to the faculty at Harvard. In 1980 he joined the faculty at the University of Illinois, moving to the University of California, San Diego, in 2000. His research has ranged widely over many areas of theoretical chemistry, physics, and biology, including theories of chemical reactions and quantum many-body phenomena in liquids and biomolecules and the theory of glasses. He is most well known, however, for his development of the energy landscape theory of protein folding, which brought the perspective of modern statistical mechanics to this central problem of molecular biology and led to new approaches to predicting protein structures from DNA sequences. Dr. Wolynes is a member of the National Academy of Sciences (1991) and a fellow of the APS, the Biophysical Society, the American Society for the Advancement of Science, and the American Academy of Arts and Sciences. He has received the Award in Pure Chemistry (1986) and the Peter Debye Award in Physical Chemistry (2000), both from the American Chemical Society. He is the 2004 recipient of the Biological Physics Prize awarded by the APS.

Robert H. Austin (NAS) is a professor of biophysics at Princeton University whose current research involves topics such as DNA conductance and nanofabrication. He received his Ph.D. in physics from the University of Illinois at Urbana-Champaign in 1975. Since 1979, Dr. Austin has been a professor at Princeton University. In 1998 he was elected as a fellow of the American Physical Society and as a fellow of the American Association for the Advancement of Science. In the following year, Dr. Austin was elected a member of the National Academy of Sciences for his ability to combine physical tools and theories with biochemical techniques to attack fundamental problems in protein and nucleic acid dynamics and function. His research interests span three areas: protein dynamics and conformational statistics; DNA dynamics and base pair sequence elastic variability; and applications of micro- and nanofabrication technology to cellular and molecular biology. He has also served as chair of the Division of Biological Physics in the American Physical Society (2002). Currently, he is the chair of the U.S. Liaison Committee for the International Union of Pure and Applied Physics.

Bonnie Bassler (NAS) is professor and director of graduate studies in the Department of Molecular Biology at Princeton University and has been an HHMI investigator since 2005. She received her Ph.D. in biochemistry from the Johns Hopkins University in 1984. Before becoming a professor, Dr. Bassler was a postdoctoral fellow and research scientist at the Agouron Institute. Her current research interests include the molecular mechanisms that bacteria use to communicate with one another. She is a fellow of the American Academy of Microbiology (2002) and the American Association for the Advancement of Science (2006). She is also a member of the National Academy of Sciences (2006). Dr. Bassler is the recipient of several awards, including the MacArthur Foundation Fellowship (2002), the Theobald Smith Society Waksman Award (2003), the Thomas Alva Edison Patent Award for Medical Technology, and the Eli Lilly and Company Research Award (2006). She was also chosen as the 2004 Inventor of the Year by the New York Intellectual Property Law Association. She is a member of several professional societies and has served on several committees and scientific advisory boards such as that for the Max Planck Institute for Infection Biology in Berlin.

Charles R. Cantor (NAS) is a founder, chief scientific officer, and member of the board of directors of Sequenom, Inc. He is also a founder of SelectX Pharmaceuticals, a drug discovery company based in the Boston area. Dr. Cantor is codirector of the Center for Advanced Biotechnology and professor of biomedical engineering at Boston University and has held positions at Columbia University and the University of California at Berkeley. He was also director of the Human Genome Center of the Department of Energy at Lawrence Berkeley Laboratory. He received his Ph.D. in biophysical chemistry from the University of California, Berkeley, in 1966. His research interests include human genome analysis, molecular genetics, new biophysical tools and methodologies, and genetic engineering. He has published more than 400 peer-reviewed articles, has been granted more than 60 patents, and coauthored a three-volume textbook on biophysical chemistry and the first textbook on genomics: *The Science and Technology of the Human Genome Project*. He sits on the advisory boards of more than 20 national and international organizations.

William F. Carroll is vice president for chlorovinyl issues of the Occidental Chemical Corporation, adjunct professor of chemistry at Indiana University, where he teaches polymer chemistry, and past president of the American Chemical Society (2006). He received his Ph.D. in organic chemistry from Indiana University, Bloomington, in 1978. Dr. Carroll is a fellow of the Royal Society of Chemistry and a member of the Science Advisory Board for DePauw University. He has been an active member of and chaired various committees for a number of chemistry, plastics, fire protection, and recycling organizations. He has served on expert groups commissioned by the United Nations Environmental Program, the State of Florida, and the Oregon Department of Environmental Quality. Dr. Carroll holds two patents and has over 40 publications in the fields of organic electrochemistry, polymer chemistry, combustion chemistry and physics, incineration, plastics recycling, and chlorine issues. He received the Vinyl Institute's Roy T. Gottesman Leadership Award for lifetime achievement in 2000. Currently, he is serving on the Chemical Sciences Roundtable of the NRC and the U.S. National Committee for the International Union of Pure and Applied Chemistry.

Thomas R. Cech (NAS, IOM) is Distinguished Professor of Chemistry and Biochemistry at the University of Colorado at Boulder and director of the Colorado Initiative in Molecular Biotechnology. Previously, he was president of the Howard Hughes Medical Institute. Dr. Cech received a B.A. degree in chemistry from Grinnell College and a Ph.D. degree in chemistry from the University of California at Berkeley. His postdoctoral work in biology was conducted at the Massachusetts Institute of Technology. Dr. Cech is a strong advocate for science education at all levels and has worked to improve the career development and mentorship of young scientists. Dr. Cech is an elected member of the National Academy of Sciences, the Institute of Medicine, and the American Academy of Arts and Sciences. Among the honors he has received are the Lasker Award, the National Medal of Science, and the 1989 Nobel prize in chemistry.

Christopher B. Field (NAS) is the founding director of the Carnegie Institution's Department of Global Ecology, where his research emphasizes ecological contributions across the range of Earth-science disciplines. Dr. Field and his colleagues have developed diverse approaches to quantifying large-scale ecosystem processes, using satellites, atmospheric data, models, and census data. At the ecosystem-scale, he has, for more than a decade, led major experiments on grassland responses to global change, experiments that integrate approaches from molecular biology to remote sensing. Dr. Field's activities in building the culture of global ecology include service on many national and international committees, including committees of the National Research Council, the International Geosphere-Biosphere Programme, and the Earth System Science Partnership. He is a fellow of the European Space Agency Aldo Leopold Leadership Program and a member of the National Academy of Sciences. He has served on the editorial boards of *Ecology*, *Ecological Applications*, *Ecosystems*, *Global Change Biology*, and *PNAS*. Dr. Field received his Ph.D. from Stanford in 1981 and has been at the Carnegie Institution since 1984.

Graham R. Fleming (NAS) is the deputy director of Lawrence Berkeley National Laboratory (LBNL) and a professor in the chemistry department at the University of California at Berkeley. He received a Ph.D. in physical chemistry from the University of London in 1974 and following several postdoctoral positions, spent 18 years at the University of Chicago. He moved to Berkeley in 1997 to direct the newly created physical bioscience division at LBNL. His research expertise is in the application of femtosecond spectroscopy to chemical and biological phenomena, recently focusing on the energy transfer steps of photosynthesis. He has also studied the difference between natural and man-made solar energy conversion materials. He served on the Chemistry Advisory Committee for the National Science Foundation and recently chaired the Grand Challenges subcommittee for the Department of Energy's Basic Energy Sciences Advisory Committee. Dr. Fleming is a fellow of the American Association for the Advancement of Science, a member of the National Academy of Sciences, and a past winner of the prestigious Guggenheim Fellowship and A.P. Sloan Foundation Fellowship awards. He has won numerous awards from the American Chemical Society, including the Nobel Laureate Signature Award for Graduate Education in Chemistry, the Peter Debye Award in Physical Chemistry, and the Harrison Howe Award.

Robert J. Full is professor of integrative biology at the University of California at Berkeley. He received his Ph.D. from SUNY Buffalo in 1984 and then held a research and teaching postdoctoral position at the University of Chicago from 1984 to 1986. In 1986 he joined the faculty of the University of California at Berkeley as an assistant professor of zoology. He was promoted to associate professor of integrative biology in 1991 and became a full professor in 1995. In 1990, Dr. Full received a National Science Foundation Presidential Young Investigators Award and in 1996 was given a Distinguished Teaching Award. In 1997, Professor Full became a Chancellor's Professor and the director of a new biological visualization center. He directs the Poly-P.E.D.A.L. Laboratory, which studies the performance, energetics, and dynamics of animal locomotion (P.E.D.A.L.) in many-footed creatures.

Shirley Ann Jackson (NAE) is the 18th president of Rensselaer Polytechnic Institute (RPI). She holds a Ph.D. in theoretical elementary particle physics from the Massachusetts Institute of Technology (MIT) (1973) and was the first African American female to receive a Ph.D. from MIT in any subject. She specializes in theoretical condensed matter physics, especially layered systems, and the physics of optoelectronic materials. Before becoming president at RPI, Jackson was chair of the U.S. Nuclear Regulatory Commission (1995-1999), where she was its principal executive officer; a theoretical physicist conducting basic research at the former AT&T Bell Laboratories (1976-1991); and a professor of theoretical physics at Rutgers University (1991-1995). Professor Jackson is a member of the National Academy of Engineering (2001) and is on the Division on Earth and Life Studies Division Committee in the NRC.

Laura L. Kiessling (NAS) is a professor of biochemistry at the University of Wisconsin-Madison and a MacArthur Foundation Fellow. She received a Ph.D. from Yale University in 1989. Her research specializes in biological recognition processes and chemical synthesis. She has attained several awards, including the Tetrahedron Young Investigator Award in Bioorganic or Medicinal Chemistry (2005). Dr. Kiessling was elected a member of the National Academy of Sciences in May 2007 and has served as the editor of *ACS Chemical Biology*.

Charles M. Lovett, Jr., is the Philip and Dorothy Schein Professor of Chemistry at Williams College. He is also director of the Science Center, chair of the Science Executive Committee and of the Bioinformatics, Genomics and Proteomics Project (BIG P). His research focuses on damage-inducible DNA repair in *Bacillus subtilis*. He has isolated and characterized many of the molecular components (i.e., regulatory proteins and DNA binding sites) of this induction pathway.

Dianne Newman is the John & Dorothy Wilson Professor of Biology and Geobiology in the Departments of Biology and Earth and Planetary Sciences at the Massachusetts Institute of Technology (MIT). She completed a Ph.D. at MIT in civil and environmental engineering, followed by postdoctoral research at the

Harvard Medical School. She spent 7 years as a professor of geobiology at the California Institute of Technology and was a researcher with the Howard Hughes Medical Institute prior to returning to MIT. Her research expertise is in molecular geobiology, using interdisciplinary approaches to study the molecular mechanisms that underlie ancient forms of metabolism. Dr. Newman received the David and Lucille Packard Foundation Fellowship for Science and Engineering and the Young Investigator Award from the Office of Naval Research.

Monica Olvera de la Cruz is professor of materials science and engineering, chemistry, and chemical and biological engineering at Northwestern University. She received a Ph.D. in physics from Cambridge University in 1985. She has been a visiting professor at the Service de Physique Theorique, Commissariat a l'Energie Atomique, in France, where she also held a staff scientist position (1995-1997). She was a Baetjer lecturer at Princeton in 2005. Currently, she is on the Advisory Committee for the National Science Foundation Mathematical and Physical Sciences Directorate (2005-2009) and the Solid State Science Committee for the NRC (2006-2010). Her expertise is in polymer theory, phase transformations, and polyelectrolytes. Dr. Olvera de la Cruz is a fellow of the American Physical Society. She received the Presidential Young Investigator Award of the National Science Foundation (1990-1995), the Alfred P. Sloan Fellowship (1990-1992), and the David and Lucile Packard Fellowship in Science and Engineering (1988-1994).

José N. Onuchic (NAS) is a professor at the University of California at San Diego (USCD), where he co-directs the NSF Center for Theoretical Biological Physics. His research is in the area of theoretical biophysics and chemical physics, focusing on the rational design of functional proteins using computational methods. He is a member of the Molecular Biophysics Training Grant Steering Committee at UCSD and served on UCSD's Task Force on Biological Sciences. He was awarded the Engineering Institute Prize, Sao Paulo, Brazil, in 1980 and the International Centre for Theoretical Physics Prize in honor of Professor Werner Heisenberg, Trieste, Italy, in 1988. He received his Ph.D. from the California Institute of Technology in 1987. He was named an associate member of the Academia de Ciencias do Estado de Sao Paulo, a Beckman Young Investigator, a fellow of the American Physical Society, and a senior fellow of SDSC, a national laboratory for computational science and engineering. He is a member of the NRC's Board on Physics and Astronomy and in 2006 was elected a member of the National Academy of Sciences for his work in developing the quantitative field of electron tunneling in proteins and explaining how electron transfer rates depend on protein structure.

Gregory A. Petsko (NAS, IOM) is the Gyula and Katica Tauber Professor of Biochemistry and Molecular Pharmacodynamics and the director of the Rosenstiel

Basic Medical Sciences Research Center at Brandeis University. He has developed low-temperature methods in protein crystallography and their use to study enzymatic mechanisms and has pioneered the study of protein dynamics in enzymatic reactions. For over 25 years, he has worked to understand how enzymes achieve their extraordinary catalytic power, developing crystallographic methods for direct observation of productive enzyme-substrate and enzyme-intermediate complexes that led to techniques for studying protein crystal structures at very low temperatures. He is a founding scientist of the combinatorial-chemistry company ArQule, Inc., and hopes to use genetic, biochemical, and biophysical tools to study structure-function relationships as they apply to in vivo and in vitro function. He was elected to membership in the National Academy of Sciences in 1995 and to the Institute of Medicine in 2001.

Astrid Prinz is an assistant professor of biology at Emory University. She earned a Ph.D. from the Munich Technical University in 2000. Dr. Prinz specializes in neural networks, most recently the stomatogastric ganglion in crustaceans, and is a member of the Computational and Life Sciences Initiative at Emory.

Charles V. Shank (NAS, NAE) is currently a member of the Janelia Farm Research group for HHMI. He received a Ph.D. in electrical engineering from the University of California, Berkeley, in 1969. He then spent 20 years as a researcher and director at AT&T Bell Laboratories in New Jersey. In 1989, Dr. Shank moved to the Lawrence Berkeley National Laboratory in Berkeley, California, where he served as director until 2004. In addition to his position as laboratory director, Dr. Shank had a triple appointment as professor at the University of California, Berkeley, in the departments of physics, chemistry, and electrical engineering and computer sciences. Dr. Shank has served on numerous state and national committees and councils, including the California Council on Science and Technology; the National Critical Technologies Panel of the U.S. Office of Science and Technology Policy, and the Solid State Sciences Committee of the National Research Council. He was chair of the NRC's Committee on Optical Science and Engineering. He has been honored with the R.W. Wood Prize of the Optical Society of America, has received the George E. Pake Prize and the Arthur L. Schawlow Prize of the American Physical Society, and is a member of the National Academy of Sciences (1984) and National Academy of Engineering (1983). He is author or coauthor of more than 200 scientific publications. His research expertise includes electro-optical systems, laser systems, and solid state electronics. Currently, Dr. Shank is a member of the Board on Physics and Astronomy of the NRC.

Boris I. Shraiman is a permanent member of the Kavli Institute of Theoretical Physics at the University of California at Santa Barbara. His research focuses on quantitative systems biology and bioinformatics and the statistical mechanics of nonequilibrium systems focusing on physical mechanisms of growth control in the development of limbs and organs and physical approaches to comparative genomics and evolution. Before coming to the Kavli Institute, he worked at Lucent Technologies and Rutgers University. He received a Ph.D. in physics from Harvard in 1983 and did postdoctoral work at the University of Chicago.

H. Eugene Stanley (NAS) is university professor, professor of physics, physiology, and biomedical engineering, and director of the Center for Polymer Studies at Boston University. He has made fundamental discoveries in the theory of phase transitions and critical phenomena for a wide range of systems, including the water structure and polymers. His pioneering applications of statistical mechanics to biology, economics, and medicine have led to significant insights. He is an elected member of the NAS.

George M. Whitesides (NAS, NAE) is the Woodford L. and Ann A. Flowers University Professor at Harvard. He received a Ph.D. from the California Institute of Technology in 1964 and was a member of the faculty of the Massachusetts Institute of Technology from 1963 to 1982. Dr. Whitesides joined the Department of Chemistry of Harvard University in 1982 and was Department Chairman from 1986 to 1989 and Mallinckrodt Professor of Chemistry from 1982 to 2004. His present research interests include physical and organic chemistry, materials science, biophysics, complexity, surface science, microfluidics, self-assembly, micro- and nanotechnology, science for developing economies, origin of life, and cell-surface biochemistry. Dr. Whitesides's recent advisory positions include service with the NRC on various boards, the National Science Foundation, the Department of Defense, and NASA. He is currently on the Committee on Science, Engineering, and Public Policy for the National Academies. He is a member of several societies, including the National Academy of Sciences (1978), the American Academy of Arts and Sciences, the National Academy of Engineering (2005), and the Royal Netherlands Academy of Arts and Sciences. He is also an honorary member of the Materials Research Society of India and honorary fellow of the Chemical Research Society of India. Dr. Whitesides has won several awards, including the National Medal of Science and the Linus Pauling Medal.